LABORATORY MANUAL

for Rost, Barbour, Stocking, and Murphy's

PLANT BIOLOGY

Deborah K. Canington

University of California, Davis

Wadsworth Publishing Company

I(T)P® An International Thomson Publishing Company

Belmont, CA • Albany, NY • Bonn • Boston • Cincinnati • Detroit • Johannesburg • London
Madrid • Melbourne • Mexico City • New York • Paris • Singapore • Tokyo • Toronto • Washington

Biology Publisher: Jack Carey
Project Development Editor: Kristin Milotich
Editorial Assistant: Michael Burgreen
Project Editor: John Walker
Print Buyer: Stacey Weinberger
Copyeditor: Dave Rich, Edit San Jose
Production: Dovetail Publishing Services
Text Design: Dovetail Publishing Services
Illustrations: Illustrious, Precision Graphics, Page Two, Maddock Illustration
Cover Design: Gary Head
Printer: Courier

Printed in the United States of America
    3   4   5   6   7   8   9   10

For more information, contact Wadsworth Publishing Company, 10 Davis Drive, Belmont, CA 94002,
or electronically at http://www.thomson.com/wadsworth.html

International Thomson Publishing Europe
Berkshire House 168-173
High Holborn
London, WC1V 7AA, England

Thomas Nelson Australia
102 Dodds Street
South Melbourne 3205
Victoria, Australia

Nelson Canada
1120 Birchmount Road
Scarborough, Ontario
Canada M1K 5G4

International Thomson Publishing GmbH
Königswinterer Strasse 418
53227 Bonn, Germany

International Thomson Editores
Campos Eliseos 385, Piso 7
Col. Polanco
11560 México D.F. México

International Thomson Publishing Asia
221 Henderson Road
#05-10 Henderson Building
Singapore 0315

International Thomson Publishing Japan
Hirakawacho Kyowa Building, 3F
2-2-1- Hirakawacho
Chiyoda-ku, Tokyo 102, Japan

International Thomson Publishing Southern Africa
Building 18, Constantia Park
240 Old Pretoria Road
Halfway House, 1685 South Africa

ISBN 0-534-24952-3

TABLE OF CONTENTS

PREFACE

The scientific study of plants is an essential component of any course in introductory biology, and laboratory work is an important part of such an endeavor. This laboratory manual is specifically designed to accompany the text *Plant Biology* by Thomas Rost, Michael Barbour, C. Ralph Stocking, and Terence Murphy. However, the manual can be used with any introductory plant biology text in conjunction with quarter- or semester-length courses.

The manual is divided into 22 exercises. Because flowering plants are our most familiar plants, the manual begins with general morphological features of the flowering plant body. Exercises 2 through 8 continue the study of flowering plants by expanding the aspects of morphology, and by introducing plant anatomy and physiology.

Exercise 9 can be used as an introduction or a review of meiosis and the general life cycles encountered in plant biology. This exercise sets the stage for the other exercises that cover flowering and nonflowering plant sexual reproduction.

Exercise 10 returns to flowering plants, introducing students to the structure and diversity of the flower. Exercises 11 through 13 continue the topic of sexual reproduction in flowering plants. Exercise 14, plant growth and development, concludes the sequence of flowering plant exercises.

Exercises 15 through 17 introduce bacteria and cyanobacteria; fungi and funguslike organisms; and algae and lichens. These three exercises are the first of seven exercises that explore the diversity of organisms. Exercises 18 through 21 continue to cover diversity within the plant kingdom with discussion of nonvascular plants, seedless vascular plants, and nonflowering and flowering seed plants.

Exercise 22 is an experiment in intraspecific plant competition that is begun early in the course and carried out over several weeks. It requires student work in two separate laboratory meetings. In addition, the exercise includes instructions and templates for student preparation of a formal experiment report.

Exercises 1 through 21 are designed in sections that allow flexibility for different academic calendars or for two- or three-hour laboratories. Although the exercise sections are intended for study in the order presented, not all sections are necessary to introduce a single botanical topic, and many of the sections can be combined into exercises occupying a single laboratory meeting. Exercises can also be expanded with supplemental material of interest to individual instructors.

Exercises begin with an objectives list and relevant terminology. Each exercise has at least two parts consisting of an introduction, a list of required materials, and instructions on how to carry out the procedure. Figures and tables are included with many of the exercises. Exercises end with student review questions. At the back of each exercise section, there is a hand-in laboratory quiz that instructors can require as part of the course, or the quizzes can be used as supplemental review material.

Encourage students to read the entire laboratory exercise before attending the laboratory session. At a minimum, student familiarity with the objective and terminology lists at the start of each exercise will assist them in learning from their laboratory experience.

Deborah K. Canington
University of California, Davis

INTRODUCTION TO THE FLOWERING PLANT BODY

OBJECTIVES

1. Describe and compare dicot and monocot seeds.

2. Describe the changes in morphology as a plant develops from the embryo within a germinating seed to a seedling.

3. Define and identify typical roots, stems, and leaves.

4. If given a modified root, stem, or leaf, be able to recognize it as a root, stem, or leaf.

5. Describe the various functions of modified vegetative plant organs.

TERMINOLOGY

adventitious bud	meristem
adventitious root	monocot
axillary bud	node
blade	petiole
branch	petiolule
bulb	plumule
coleoptile	primary root
coleorhiza	rachis
compound leaf	radicle
corm	rhizome
cotyledon	root
dicot	root system
embryo	scutellum
endosperm	seed
epicotyl	seed coat
fibrous root	seedling
grain	shoot apex
herbaceous plant	shoot system
hilum	shoot tip
hypocotyl	simple leaf
internode	spine
lateral root	stem
leaf	stipule
leaf primordium	stolon
leaf sheath	taproot
leaf vein	terminal bud
leaflet	tuber

THE VEGETATIVE BODY OF THE FLOWERING PLANT

The flowering plant body has three vegetative organs: **stems**, **leaves**, and **roots**. (Flowers are organs of sexual reproduction.) Despite the simplicity of having only three vegetative organs to sustain their existence, plants evolved endless variations of these organs, revealing adaptations to many types of habitats. Within the Magnoliophyta, the flowering plant division, there are two classes: the Magnoliopsida (informally called the dicotyledones or **dicots**) and the Liliopsida (or **monocots**). Today's exercise introduces some of the morphological and anatomical differences between these classes, and additional differences will be revealed throughout the course. Our plant study begins with comparing the seeds and seedlings of two familiar dicot and monocot plants: bean and corn.

PART A   SEED TO SEEDLING

Flowering plant **seeds** contain an **embryo** and a food supply, all surrounded by a protective **seed coat**. Depending on the species, the food supply may be stored in enlarged **cotyledons** or in a nonembryonic tissue called **endosperm**. When conditions are right, the seed imbibes water and germinates. The young root, called a **radicle**, is the first organ to emerge through the seed coat. A **seedling** is a young, immature plant recently germinated from a seed. In this portion of the exercise, you will compare seeds and seedlings of dicot and monocot plants.

MATERIALS

- soaked bean seeds
- two-week-old bean seedlings
- soaked corn grains
- two-week-old corn seedlings
- I$_2$KI stain
- dissecting microscope
- single-edge razor blades
- dissecting needle

PROCEDURE

1. A Typical Dicotyledon Seed and Seedling

   a. The bean seed

   Pick up a soaked bean seed (*Phaseolus vulgaris*) and examine its external morphology. Visible along one long edge of the seed is the **hilum**, a small elongated scar left after the seed detaches from its stalk. The presence of two large, pale cotyledons, the most conspicuous part of the bean embryo, identifies the bean as a dicotyledonous plant. The cotyledons are seed leaves borne by the young embryo axis. Separate the cotyledons, and look for the shoot-root axis between the them. The embryonic shoot is the **plumule**, and the embryonic root is the radicle. The plumule in a bean has small, unexpanded foliage leaves, and the radicle is the first organ to emerge through the seed coat. Look among the soaked bean seeds for ones that show signs of germination.

   b. Organs of the bean seedling

   Examine a two-week-old bean seedling. During germination the radicle emerges and lengthens, becoming the **primary root**. Look for the remains of the primary root and for **lateral roots** emerging from other roots. The stem is that portion of the plant axis segmented into **nodes** and **internodes**. The nodes are regions where lateral appendages, such as leaves, arise. The internodes are the regions of stem between the nodes. Internodes below and above the cotyledonary node have special names: the **hypocotyl** and the **epicotyl**, respectively.

   Axillary buds are located at nodes and in the axil between a foliage leaf and the stem axis. Axillary buds grow into lateral shoots, or **branches**. Bean plants produce **simple foliage leaves** and **compound foliage leaves**. Simple leaves have a stalk (the **petiole**) and a **blade**. Compound leaves have a petiole, and the blade is divided into **leaflets**, each with its own **petiolule**. The extension of the petiole to which leaflets are attached has the special designation of **rachis**. Pairs of small leaf-like appendages called **stipules** may be present at the base of each petiole. Identify nodes, internodes, cotyledons, simple leaves, compound leaves, axillary buds, and roots.

   c. The shoot tip

   The **shoot tip** is at the end of the bean stem, but it does not appear to be the highest part of the bean plant; young leaves extend above it, and the petioles can be mistaken for internodes. To find the shoot tip, work your way up the seedling, moving from one internode to the next. Each node has a petiole, an axillary bud, an internode that arrives from below, and an internode that extends upward. Follow the internode to the next node. Eventually, you will reach a point where the leaves are so tiny and the internodes are so short that a dissecting microscope is needed to distinguish them. At this stage, leaves are referred to as **leaf primordia**.

   This region, with several leaf primordia and the end of the stem, is the shoot tip. Examine the shoot tip with the aid of a dissecting microscope. With a probe, gently move the leaf primordia aside to expose the **shoot apex** of the shoot tip. The shoot apex is the dome-shaped summit of the shoot tip, the site of the apical **meristem**, the source of new stem cells, and it is the **terminal bud**.

2. A Typical Monocotyledon Seed and Seedling

   a. The corn grain

   Corn **grains** (*Zea mays*) are fruits containing a single seed where the seed coat and fruit wall are fused. (Fruit types will be covered in a later exercise.) In corn and other grasses, the seed's food reserves are not stored in cotyledons. Instead they are stored beside the embryo in a tissue called endosperm. This is typical of monocotyledons as is the presence of a single cotyledon. The single cotyledon of corn, called the **scutellum**, is a narrow shield of tissue that is pressed against the endosperm. (There will be more information on endosperm in later exercises.) Grass seeds have special tissues that protect the embryo shoot tip and the radicle. The **coleoptile** is a sheath of protective tissue covering the shoot tip, and the **coleorhiza** is a protective sheath of tissue surrounding the radicle.

   Examine a soaked corn grain. The embryo is visible on one side of the grain as light-colored tissue with a pointed end. To examine the internal structure of the corn grain, lay the grain on a cutting surface with the embryo facing up. Hold a single-edge razor blade perpendicular to the cutting surface, and slice the grain in half, bisecting the embryo axis. Applying a drop of I_2KI solution to the cut surface of the grain will stain starch a dark blue or black, making it easier to distinguish between embryo and food reserves. Allowing the cut surface to dry for a few minutes may improve visibility of the separate tissues.

b. Organs of the corn seedling

Examine a two-week-old corn seedling. First locate the remnants of the grain, which can be found among the roots. Then locate the large primary root with its lateral roots. Note the stem structure emerging from the grain. The first portion of the stem visible above the grain is an internode, then a node is recognizable as a thickening of the stem. The first green leaf is attached at the node. **Adventitious roots** (roots that arise from an organ other than another root) also emerge from this node. Carefully peel off the outermost leaf, noting the long **leaf sheath** that extends upward from the node where the leaf emerges from the node. The leaf sheath wraps around the stem. To estimate the length of the stem in a corn seedling, peel off the outer leaves, taking care not to break off the delicate inner organs enclosed by the leaves, and expose the nodes and internodes.

c. The shoot tip

Rather than being exposed at the top of the plant as in the bean, the shoot tip in a young grass such as the corn seedling is enclosed within many leaf sheaths. To find the tip, cut off the shoot about 2 cm above the first node. Then, with a sharp single-edge razor blade, slice the shoot in half lengthwise. Inspect both halves with a dissecting microscope, and use the half that seems likely to contain the shoot tip. Using the dissecting microscope, note that leaf primordia arise from a pale, conical stem. The shoot apex is at the tip of the cone. To expose the apex, peel away the leaves with a dissecting needle.

PART B MORPHOLOGY OF MATURE VEGETATIVE PLANT ORGANS

Stems and leaves form the **shoot system**, and the primary root and lateral roots form the **root system**. Although shoots normally define the aerial portion of the plant axis and roots the subterranean portion, modifications of stems, leaves, and roots are plentiful. In this exercise, typical plant organs and examples of modifications of plant organs are provided for study. When observing a modified organ, compare it with the typical organ of the same type.

MATERIALS

- plant with taproot system
- plant with fibrous root system
- carrot root
- ivy with aerial roots
- white potato tuber
- iris rhizome
- ginger rhizome
- strawberry stolon
- gladiolus corm
- onion bulb
- celery petiole
- insectivorous plants
- *Bryophyllum* with foliar plantlets
- succulent plant with fleshy leaves
- cactus plant
- *Coleus* plant
- leaf series: simple and compound pinnate and palmate leaves
- monocot leaves

PROCEDURE

1. Roots

   a. Taproot and fibrous root systems

   There are two basic types of root systems, **taproot** and **fibrous**. Taproots are enlarged primary roots. In plants with fibrous root systems, the primary root may form many lateral roots of equal size, or the root system may consist of many adventitous roots. Examine the living root systems on display. Try to identify the origin of the fine roots. Are they adventitious roots or lateral roots?

   b. Modified roots

   Examine the display of modified roots on the side bench. While examining the specimens, try to deduce the functions of the modified roots from their appearance and compare them with the fibrous and taproots seen earlier in the lab.

2. Stems

   a. Herbaceous stems

   Bean and corn plants are **herbaceous plants**; that is, they lack woody growth. Large trees, such as oaks and maples, are examples of plants with woody growth. (Woody plants will be examined in a later lab.) Again, compare the stems of the bean and corn seedlings with the stems of older bean and corn plants displayed on the side bench. Use these plants as a reference when examining the modified stems.

b. Modified stems

Attempt to identify the modified organs as stems, remembering that stems have nodes. You may recognize nodes by the presence of leaves or axillary buds, or by a scar that rings the stem where a leaf was attached. In each case, decide what has been modified and on the major function of the modified stem. Differentiate among **stolons**, **rhizomes**, **corms**, and **tubers**.

3. Leaves

a. Simple to compound leaves

Since plants live in a myriad of environments, and leaves may perform a variety of functions, leaf morphology is highly variable. Sessile (lacking petioles), petiolate (petioles present), simple, and compound leaves are on display, as are leaves with parallel or netted venation. Examine the overall morphology of the specimens and examine the venation of the leaves. (Venation is the arrangement of veins in a leaf blade.) Note if there is more than one major **leaf vein** and the branching of minor veins. To distinguish between a petiole and a stem, look for buds or shoot tips. Stems have axillary buds or shoot tips along the stem axis and a terminal bud or shoot tip at the end of the axis; leaves do not. If you think an organ is a leaf, look for an axillary bud or axillary shoot at the junction of petiole and stem axis. You will not find axillary buds at the base of the petiolule of a leaflet of a compound leaf. This can be useful in determining whether a specimen is a compound leaf or whether it is a stem with simple leaves.

b. Dicot and monocot leaves

Examine the specimens indicated as representative dicot and monocot leaves. What is the major difference between the major leaf vein patterns in typical dicot and monocot leaves? Are there other differences in leaf morphology?

c. Modified leaves

Examine the specimens on display. Note that onion **bulbs** have two different types of modified leaves. Can you find any evidence indicating that cactus **spines** are modified leaves? Why are buds on leaf blades called **adventitious buds**? Compare the modified leaves with the typical leaves, and use your imagination and logic to decide on possible functions of the modified leaves or leaf parts.

QUESTIONS FOR THOUGHT AND REVIEW

1. What is the major difference between a dicot and a monocot? _____

_____

2. Where is food stored in the mature bean seed?

_____

3. Where would you look for a coleoptile in a corn grain? _____

4. What organs make up the vertical axis of a young corn plant? _____

_____

5. How can you differentiate between a simple leaf and a compound leaf? _____

_____

6. What type of root system would you expect in plants growing in regions with a very low water table and infrequent but heavy rains? Why? _____

_____

_____

_____

7. Other than the number of cotyledons, list several differences in the vegetative morphology of dicot and monocot plants. _____

_____

_____

_____

8. Compare and contrast a bulb and a corm. _____

_____

_____

9. Name two types of modified horizontal stems.

_____

_____

10. Describe one type of modified leaf and list the advantages of the modifications for the plant. _____

_____

_____

_____

Name:_____

Section Number: _____

INTRODUCTION TO THE FLOWERING PLANT BODY

1. List the three parts of a flowering plant seed.

    a. _____

    b. _____

    c. _____

2. From which part of a seed do cotyledons arise? _____

3. What is the difference between the hypocotyl and the epicotyl? _____

    _____

    _____

4. Contrast the location of the shoot apical meristems in bean seedlings and in corn seedlings. _____

    _____

    _____

5. List the three vegetative organs of flowering plants.

    a. _____

    b. _____

    c. _____

6. A rhizome is a modified _____.

7. Each fleshy component of an onion bulb is a modified _____.

8. What is the function of a fleshy leaf on a succulent plant?_____

    _____

EXERCISE 2

VIEWING PLANT CELLS

OBJECTIVES

1. Learn the correct procedure for handling, adjusting, and using a compound microscope.

2. Prepare fresh wet mounts of plant tissue for observing with a compound microscope.

3. Identify plant cell walls, plasmodesmata, nuclei, chloroplasts, chromoplasts, amyloplasts, and starch grains.

4. Distinguish between turgid and plasmolysized plant cells.

5. Identify cells in the different stages of mitosis.

6. Describe the major event of cytokinesis.

TERMINOLOGY

amyloplast	metaphase
anaphase	middle lamella
anthocyanin	mitosis
Brownian movement	nuclear envelope
carotenoid	nucleolus
cell	nucleus
cell plate	objective lens
cell wall	ocular lens
chlorophyll	organelle
chloroplast	osmosis
chromatid	plasma membrane
chromoplast	plasmodesma
chromosome	plasmolysis
condenser	plastid
cytokinesis	prophase
cytoplasm	protoplast
cytoplasmic streaming	spindle apparatus
diffusion	telophase
interphase	turgid
iris diaphragm	turgor pressure
kinetochore	vacuole

INTRODUCTION TO THE PLANT CELL

PART A HANDLING THE COMPOUND MICROSCOPE

A compound microscope is composed of two systems of lenses: the **objective**, nearest the specimen being viewed, and the **ocular**, nearest the eye of the viewer. Compound microscopes are useful for viewing thin cells or tissues by the light transmitted through the tissues. The basic compound microscope has brightfield optics. Brightfield means that light passes unchanged through the lens system, producing a brightly lighted field for viewing specimens.

The compound microscope is a basic research tool of biologists, and successful work with it requires an understanding of its proper use and care. Get a compound microscope from the cabinet and take it to your work space. Always carry your microscope with two hands, one hand under the base and the other hand grasping the arm. Plug it in and turn on the illumination. In addition, pick up a microscope slide designated as a "practice" slide.

MATERIALS

- compound microscope
- prepared slide with the letter e
- glass cleaner
- cotton-tipped swabs
- lens paper

PROCEDURE

While following the instructions from your instructor, go through the following steps to adjust the oculars and set the illumination for best viewing. Poor lighting ruins learning and causes eyestrain.

1. Find the **iris diaphragm** lever on the side of the condenser, below the stage. Open the iris diaphragm all the way.

2. Slowly rotate the nose piece until the 10× objective lens clicks into place in the light path. Do not hit the stage. You may have to lower the stage to avoid contact. *CAUTION! When you change to higher power lenses, watch from the side to be sure that you do not hit the slide.*

3. Turn the light source to medium intensity.

4. Find the two focus knobs below the stage. The large outer knob is for coarse focus; *use it only with the 4× and 10× objectives.* The small inner knob is the fine focus control. Using the coarse focus knob, focus on the specimen. When the specimen image begins to be sharp, continue with the fine focus knob.

5. Find the **condenser** apparatus below the stage. Using the condenser knob, move the condenser lens all the way up. Now, slowly move it down (and perhaps back up) until the image is the brightest possible. When properly adjusted, the condenser should be near its uppermost position.

6. Open and close the iris diaphragm until the image has the best contrast and detail. At this point the light from the condenser is focused on the specimen, giving a cone of light that just fills the lens. A narrower opening gives a dim image; a wider opening spreads the cone beyond the objective lens and creates glare, reducing contrast between specimen and background.

7. Adjust the eyepieces to fit your needs. First, close your left eye (or put a card between your left eye and the microscope) and adjust the fine-focus knob to get the sharpest image. Then close the right eye, open the left eye, and turn the knurled ring on the left eyepiece until the image is sharp. Finally, adjust the separation of the eyepieces to fit the spacing of your eyes. To do this, look through both eyepieces and slide them farther apart or closer together until your eyes are comfortable.

8. The microscope is now adjusted for the 10× objective. When you switch to another objective, you may have to repeat steps 5 and 6 for best lighting. *While studying objects under high power (40×) of the compound microscope, you should make a habit of keeping one hand on the fine adjustment knob and almost continually changing the focus slightly. This will enable you to visualize the positions of bodies within cells and of cells within tissues in relation to one another and to form a mental image of their three-dimensional shapes.*

Cleaning lenses

9. Use cotton-tipped swabs moistened with glass cleaner. Lens paper and glass cleaner can be used, but never use paper tissues or paper towels, which will scratch lenses. With a moistened swab or lens paper, gently wipe with a circular motion.

10. If your lenses remain dirty or are oily, ask your instructor to clean the lenses with xylene.

11. *Because many students use the same microscope each week, you will have to clean and readjust your microscope at the beginning of each lab meeting.*

Microscope storage

12. When returning the microscope to its cabinet, first remove any slide on the stage and, as a courtesy to your fellow students, return the slide to its proper tray. Coil the cord and then drape it over an eyepiece; *do not* wind the cord around the substage or objective lenses. Again, carry it with two hands. If you think your microscope is broken, attach a note to the microscope indicating what is broken and immediately tell your instructor. *DO NOT RETURN BROKEN MICROSCOPES TO THE CABINET.*

PART B   VIEWING PLANT CELLS

Cells are the basic unit of a plant, and all plants are multicellular. Before examining specific types of plant cells and how these cells are arranged in tissues and organs, this exercise explores the structure of a few typical plant cells. Plant cells consist of a **cell wall** and a **protoplast**, and the protoplast is delimited by the **plasma membrane** and contains **cytoplasm** and a **nucleus**. The nucleus is surrounded by a **nuclear envelope** and contains one or more dense regions called **nucleoli**. Located within the ground substance of the cytoplasm are membrane-bound **organelles** and systems of membranes, including nuclei, **plastids**, mitochondria, dictyosomes, endoplasmic reticulum, and **vacuoles**. Plant cells are cemented together by pectic compounds. This region of intercellular substances is the **middle lamella**. Most plant cells are cytoplasmically connected with their neighboring cells through minute pores called **plasmodesmata**. This portion of the exercise compares plant cells from different plant organs, emphasizing some of the cellular organelles contained within the various cells.

MATERIALS

- *Elodea* leaf
- *Tradescantia* flower
- marigold flower petal
- tomato fruit
- white potato tuber
- I_2KI stain
- prepared slide of *Diospyros* endosperm
- compound microscope
- single-edge razor blades
- forceps
- microscope slides
- coverslips
- dH_2O in dropper bottle

PROCEDURE

1. **The Living Plant Cell**

   a. *Elodea* leaf

   Using forceps, remove a small leaf from an *Elodea* plant and place the whole leaf in a drop of water on a microscope slide. Gently lower a coverslip over the leaf, keeping the leaf flat, and view with the 10× objective lens of a compound microscope. To reveal the three-dimensional shapes of the cells and organelles, slowly focus up and down through the tissue. Note the overall shape of the cells and the numerous green **chloroplasts** within each cell. Chloroplasts are the sites of photosynthesis and the location of **chlorophylls** and **carotenoids** which are photosynthetic pigments. In healthy cells, chloroplasts exhibit an orderly movement called **cytoplasmic streaming**. Change to high power and examine the contents of an individual cell. Look for streaming of chloroplasts, and try to determine the location of chloroplasts within the cell. Locate a nucleus. It will appear pale yellow and clear.

   b. *Tradescantia* flower

   Your instructor will point out the stamens of a *Tradescantia* flower. Using forceps, remove a single stamen from the flower and place the stamen in a drop of water on a microscope slide. Add a coverslip. Locate a multicellular hair on the surface of the stamen and focus on a single hair cell. Look for the cell vacuole, it contains a purple water-soluble pigment called **anthocyanin**. Anthocyanin is not involved in photosynthesis but is commonly found in plant organs that appear pink or purple. The pigment makes it easy to see the vacuole. Note the size and location of the vacuole within the hair cell.

   c. Marigold flower petal

   Make a wet mount of tissue torn from a marigold flower petal. Using high power of a compound microscope, locate a region that is one cell thick and examine the cell contents of a single cell. The yellow pigments are carotenoid pigments and are contained in organelles called **chromoplasts**. Compare the shape, size, and location of the marigold chromoplasts with chloroplasts in *Elodea*.

   d. Tomato fruit

   Make a wet mount of tissue from a tomato fruit by scraping tissue from the tomato pulp and adding a drop of water. Do not use the "skin" of the tomato. Locate chromoplasts containing red carotenoid pigments. Compare tomato fruit chromoplasts with those in the marigold petal.

   e. Potato tuber

   Scrape tissue from a white potato tuber, avoiding the "peel." Place the tissue in a drop of water on a microscope slide and add a coverslip. Examine the preparation on high power of a compound microscope. The tissue will appear colorless, but starch grains may be visible clustered in plastids. Nonphotosynthetic plastids containing starch grains are called **amyloplasts**. If the preparation is good, remove the slide from the microscope stage and stain the potato tissue with I_2KI. To stain, place several drops of I_2KI at one edge of the coverslip and a piece of paper towel at the opposite side of the coverslip. The paper towel will draw the stain across the tissue. Reexamine the tissue. The I_2KI solution stains starch purple or black, so starch grains should be visible within the amyloplasts.

2. **The Plant Cell Wall**

   Obtain a prepared slide of persimmon (*Diospyros*) endosperm and examine the slide on high power of a compound microscope. The cell walls of this tissue are easy to see because they are unusually thick. In addition, numerous plasmodesmata are clearly visible as lines transversing the cell walls. The plasmodesmata are living, cytoplasmic connections between adjacent cells.

PART C   OSMOSIS IN PLANT CELLS

Kinetic energy in all atoms and molecules keeps molecules within plant cells in constant motion. This motion, called **Brownian movement**, is visible as a never-ending, random movement of small particles within cells when viewed at high magnification of a light microscope. (The actual atoms and molecules are too small to be seen.) This kinetic energy of molecules is the driving force of **diffusion**. Diffusion is a spontaneous, physical process where molecules move from a region where they are in a relatively high concentration to a region where they are in a relatively low concentration. (Diffusion also refers to the movement of molecules from a region of high heat or high pressure to a region of lower heat or pressure. In most living systems, heat and pressure are relatively constant, but there are exceptions, and you should keep all three factors in mind.) Diffusion of water through a selectively permeable membrane is called **osmosis**. Therefore, in osmosis, water moves across a membrane from a region of high water concentration to a region of lower water concentration.

For proper functioning and growth, plant cells require water. A healthy plant cell continually takes in water by osmosis to keep the cell stiff, or **turgid**. The pressure that develops within a turgid plant cell is called **turgor pressure**. In a cell with a high turgor pressure, the protoplast exerts an outward force (turgor pressure) against the cell wall and the cell wall exerts a backpressure (wall pressure) against the protoplast. At this point, net water movement into the cell stops. Because plant cells are surrounded by a cell wall, plant cells never burst. On the other hand, if conditions cause a plant cell to lose so much water that turgor pressure is lost, the protoplast and vacuole will shrink and the plasma membrane will pull away from the cell wall. This condition is called **plasmolysis**. Plant leaves may wilt if the leaf cells lose turgor pressure.

MATERIALS

- *Elodea* leaf
- 0.6M sucrose solution in dropper bottle
- dH$_2$O in dropper bottle
- compound microscope
- forceps
- microscope slides
- coverslips

PROCEDURE

Remove a healthy young leaf from an *Elodea* plant and make a wet mount using a 0.6M sucrose solution. After adding a coverslip, examine the tissue with the high power of a compound microscope. Note how the protoplast is pulled away from the cell wall. If 50% of the cells within the tissue are in this condition, the tissue is said to be plasmolyzed. If plasmolysis is not too severe or has not lasted too long, it can be reversed. To reverse plasmolysis, remove the slide from the microscope stage and flush pure water under the coverslip (add water from a dropper bottle to one side of the coverslip, while holding paper toweling at the opposite side). Once the sucrose solution is removed and replaced with pure water, examine the preparation again. Look for cells that have recovered turgor pressure.

PART D MITOSIS AND CYTOKINESIS IN PLANT CELLS

In order to increase the number of cells in a tissue, plant cells must divide. Cell division is one of four phases in the plant cell cycle. Three of the cell cycle phases, G1, S, and G2, are collectively called **interphase**. It is during interphase that the cell duplicates its DNA in preparation for cell division. The fourth phase of the cell cycle, cell division, has two distinct stages: **mitosis** and **cytokinesis**. Mitosis is division of the cell nucleus, and cytokinesis is division of the cytoplasm.

Mitosis progresses through four stages (Figure 2-1), beginning with **prophase**, continuing through **metaphase** and **anaphase**, and ending with **telophase**. During prophase the DNA molecules thicken to form dense **chromosomes** visible as two **chromatids** connected by a **kinetochore**, and the **nuclear envelope** breaks down. During metaphase the chromosomes are aligned along the equatorial plane of the cell, and spindle fibers extend from each kinetochore to the cell poles. The **spindle apparatus** is involved in moving the chromatids. The cell quickly moves to the next

Figure 2-1
Stages of mitosis: prophase, metaphase, anaphase, and telophase. Note movement of chromosomes and formation of cell plate.

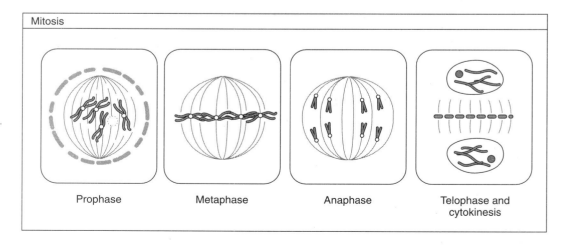

Mitosis

Prophase   Metaphase   Anaphase   Telophase and cytokinesis

stage, anaphase. During anaphase the paired chromatids separate and move toward opposite poles of the cell. Telophase marks the stage when chromatids have reached their respective poles and the nuclear envelope reforms. During telophase the chromosomes begin to uncoil, and at the end of telophase the cells are ready to begin the G1 phase of the next cell cycle.

Not all plant cells undergo cytokinesis immediately following mitosis. If cytokinesis does not occur, the result is a multinucleate cell. Usually, cytokinesis does occur, and it begins during telophase. During cytokinesis new cell walls and a middle lamella are formed. The developing cell walls and middle lamella, called a **cell plate**, are deposited from the center of the cell outward, finally reaching the side wall of the parent cell. The process of mitosis and cytokinesis results in two new cells that are identical to the parent cell. Onion root tips will be used to study the stages of mitosis.

MATERIALS

- prepared slide of *Allium* root tip in longitudinal section
- compound microscope

PROCEDURE

1. Interphase

   Obtain an *Allium* root tip slide and use low power of a compound microscope to locate the apex of the root. Change to high power and observe individual cells within the section near the apex of the root. All of the stages of mitosis are visible. First, locate a cell in interphase, the metabolic state in which chromosomes appear as a mass of chromatin and not as individual structures. Nucleoli should appear dark and should be visible within the nucleus.

2. Prophase

   Scan the root tip for a cell in prophase. Look for distinct, connected chromatids within the clear region of the nucleus. The nuclear membrane will have disappeared by the end of prophase.

3. Metaphase

   Look for cells in metaphase. The chromosomes will be aligned along the equatorial plane of the cell. The spindle apparatus may be visible.

4. Anaphase

   Cells in anaphase reveal separated chromatids. Throughout this stage, the chromatids are visible at various stages of being pulled toward the cell poles.

5. Telophase

   During telophase, the chromosomes uncoil and the nuclear envelope reappears. Look for cells with two nuclei. At this stage cytokinesis occurs, and the newly forming cell wall will be visible.

QUESTIONS FOR THOUGHT AND REVIEW

1. What is the three-dimensional shape of the *Elodea* leaf cells and of *Elodea* leaf cell chloroplasts?

_____

_____

2. What is the location of chloroplasts within the *Elodea* leaf cells? _____

_____

3. How are chromoplasts different in shape from chloroplasts? Do chromoplasts contain chlorophylls?

_____

_____

4. How can the presence of starch in plant tissue be determined? _____

_____

5. Describe one possible significance of plasmodesmata in the function of plant tissue. _____

_____

_____

6. Define turgor pressure. _____

_____

7. Under what conditions would turgor pressure increase in a plant cell? _____

_____

8. Are plant cell plasma membranes permeable to water? _____

9. During which stage of the cell cycle is DNA replicated, and during which stage of mitosis do the chromosomes separate? _____

_____

10. What is the cell plate and when does it appear during mitosis? _____

_____

EXERCISE 2
LABORATORY QUIZ

Name:_____

Section Number: _____

VIEWING PLANT CELLS

1. What surrounds the plant cell plasma membrane? _____

2. Contrast diffusion and osmosis. _____

    _____

    _____

3. What is anthocyanin? _____

4. What is the name of the cell organelle that is the site of photosynthesis? _____

5. During which stage of mitosis would you expect to see a cell plate form?_____

6. Contrast mitosis and cytokinesis. _____

    _____

    _____

7. Chromatids are located in which region of the cell? _____

8. If a plant cell is turgid, what does this tell you about the cell's turgor pressure? _____

    _____

    _____

EXERCISE 3

PRIMARY GROWTH: PLANT CELL AND TISSUE TYPES

OBJECTIVES

1. Define plant apical meristem, name the locations within a plant body where apical meristems are located, and characterize the cell type of apical meristems.

2. List the three primary meristems, describe their origin, and describe the mature cell types produced by each.

3. Define and describe the results of primary growth in plants.

4. List cell types with primary cell walls only, and describe the features and functions of these cells.

5. List plant cell types with primary and secondary cell walls, and describe their features and functions.

6. Name the cell type that provides the outer covering of a plant and the cell types that provide internal conduction within the plant body, and describe the unique features and functions of these cell types.

TERMINOLOGY

apical meristem	primary meristem
callose	primary wall
chlorenchyma cell	procambium
collenchyma cell	protoderm
companion cell	sclereid
cortex	sclerenchyma cell
dermal tissue system	secondary wall
epidermal cell	sieve area
epidermis	sieve plate
fiber	sieve tube
ground meristem	sieve-tube member
ground tissue system	stoma
guard cell	tracheary element
P-protein	tracheid
parenchyma cell	trichome
perforation plate	vascular tissue system
phloem	vessel
pit	vessel member
pith	xylem
primary growth	

INTRODUCTION TO PLANT CELL AND TISSUE TYPES

Plants are multicellular organisms made of both mature cells and cells that remain embryonic, retaining their ability to divide. This exercise will examine different types of plant cells and the tissues they form as a prerequisite to studying the cells and tissues within the organs (root, stem, leaf, and flower) that make up the plant body. Each plant cell originates from a cluster of meristematic cells, and this exercise will explore the meristematic cells involved with **primary growth**. Subsequent exercises on roots and stems will examine additional meristematic regions.

PART A THE APICAL AND PRIMARY MERISTEMS

Cells that indefinitely retain their ability to divide are clustered in tissues called **apical meristems**, located at the tips of roots and shoots. Apical meristems give plants the potential for eternal youth. Shoot tips end in shoot apical meristems and root tips have root apical meristems. The cells of the apical meristem tissue are called meristematic cells and provide an endless supply of new cells for the shoot and root systems. Because plants grow at their tips, cells produced by apical meristems are left behind as the meristem continues to grow. The cells left behind differentiate and mature into the cells of mature plant tissues. First, the apical meristems produce the three **primary meristems**, which in turn produce all of the tissue within herbaceous (nonwoody) plants. Each primary meristem produces a specific tissue or region within the plant body. Plants with woody growth have additional meristems that produce secondary growth, which will be covered in a later exercise.

The three primary meristems are the **protoderm**, the **ground meristem**, and the **procambium**. The cells of the protoderm produce the outer covering of the plant, the **epidermis** (part of the **dermal tissue system** of mature plants). An additional dermal tissue type, the periderm, is produced by plants with secondary growth. The procambium differentiates into the conducting tissues of the plant, **xylem** and **phloem**, which form the **vascular tissue system** of mature plants. An additional meristem contributes to the vascular tissue system in plants with secondary growth. In herbaceous plants, xylem and phloem are located side by side in strands throughout the plant. The remaining tissue of a mature plant arises from the ground meristem, which produces the **ground tissue system**. In leaves and young stems, the ground tissue system forms the bulk of the organs. The prepared slides in this portion of the exercise reveal the locations and appearances of shoot apical meristems in both longitudinal and cross-sectional views and a root apical meristem in longitudinal view.

MATERIALS

- prepared slide of *Coleus* shoot tip in longitudinal section
- prepared slide of *Coleus* shoot tip in cross section
- prepared slide of *Raphanus* root tip in longitudinal section
- compound microscope

PROCEDURE

1. *Coleus* Shoot Tip in Longitudinal Section

   First, look at the shoot tip on the slide without the aid of a microscope. The tissue on the slide is a terminal bud, and the axis and maturing leaves are visible with the naked eye. View the slide on low power of a compound microscope, and locate the apex of the shoot (Figure 3-1). Change to a higher power objective lens and study the apex. Note the cluster of cells at the apex. These cells form the shoot apical meristem. Note their uniform size and cellular density. The nuclei of meristematic cells are large relative to the cell size, and the cells are intensely stained relative to surrounding cells outside of the apical meristem. Note the extent of the apical meristem within the shoot apex.

2. *Coleus* Shoot Tip in Cross Section

   Using a low power of the compound microscope, locate a cross section through a *Coleus* shoot tip. Switch to a higher power and examine the cells of the apical meristem. The meristematic cells are visible as intensely stained cells with large nuclei. Because cells of the apical meristem are isodiametric, they should appear the same shape and size as they did in longitudinal view.

3. *Raphanus* Root Tip in Longitudinal Section

   Examine the *Raphanus* root tip with a compound microscope. First, locate the root apex and then

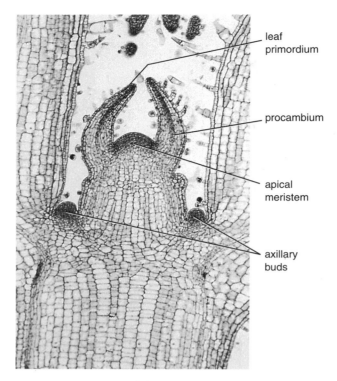

Figure 3-1 *Coleus* shoot tip, longitudinal section. The shoot apical meristem and the three types of primary meristematic tissue are visible.

examine the details of the root apical meristem. Unlike shoots, the root apical meristem is not at the extreme tip of the root apex. Roots have a protective covering capping the meristem, the root cap, and the root apical meristem will be located inside the cap. Look for cells that are isodiametric and intensely stained. Compare the root apical meristem cells and tissue with the cells and tissue of the *Coleus* shoot apical meristem.

PART B MATURE CELL AND TISSUE TYPES

In herbaceous plants, the protoderm differentiates into the epidermis. The epidermis is composed of **epidermal cells**, **trichomes** (or hairs), and **guard cells**. Epidermal cells are generally flat, platelike cells that interlock with adjacent epidermal cells. Trichomes may be unicellular or multicellular, and some are tipped with glands. Pairs of guard cells form **stomata**, controllable pores in the epidermis.

The ground meristem matures into the ground tissue system, and the most common type of cell within the ground tissue system is the **parenchyma cell** (Figure 3-2). Parenchyma cells can be large and are usually isodiametric with thin **primary walls**. A simple tissue composed of pure parenchyma cells is called

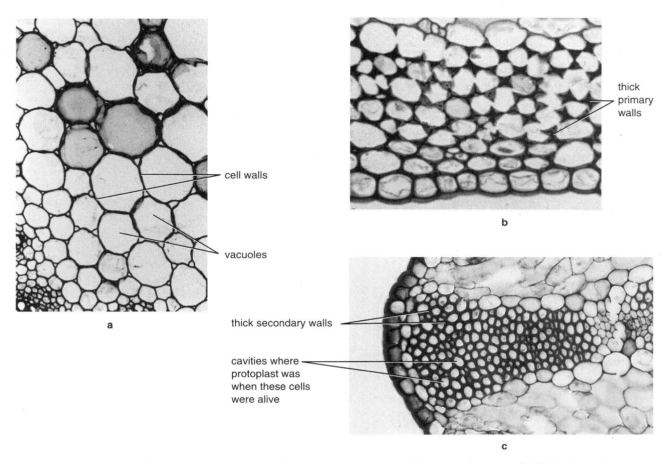

cell walls

vacuoles

a

thick primary walls

b

thick secondary walls

cavities where protoplast was when these cells were alive

c

Figure 3-2   Three types of simple, mature plant tissues in cross-sectional views. (**a**) Parenchyma tissue. (**b**) Collenchyma tissue. (**c**) Sclerenchyma tissue.

parenchyma tissue and is the common storage or photosynthetic tissue of plants. Other cell types may be present in more complex tissues, including collenchyma and chlorenchyma cells. **Chlorenchyma cells** are parenchyma cells containing chloroplasts. **Collenchyma cells** have unevenly thickened primary cell walls and collenchyma tissue is common in regions of plants requiring flexible support. A third basic cell type is the **sclerenchyma cell** and is commonly associated with vascular tissue. There are two basic types of sclerenchyma cells, **sclereids** and **fibers**. Sclereids are isodiametric in shape or they may have unusual shapes, but fibers are always elongated cells. Sclereids and fibers have primary and **secondary walls**, are lignified, and are usually dead at maturity. Lignin is extremely strong and makes cells rigid. Sclerenchyma cells provide plant tissue with strength.

The procambium matures into the vascular tissue. Vascular tissue includes xylem tissue and phloem tissue. Xylem tissue contains the plant's water-conducting cells, and phloem tissue contains the plant's sugar-conducting cells. There are two types of water-conducting cells, **tracheids** and **vessel members**. Collectively, tracheids and vessel members are called

tracheary elements, a useful term when the cells cannot be distinguished. Tracheids are elongated cells with primary and secondary cell walls that are lignified and have **pits** in the secondary walls. Vessel members are elongated cells, but wider than tracheids, with lignified primary and secondary cell walls, pits, and open-end walls called **perforation plates**. The presence of the perforation plate distinguishes a vessel member from a tracheid. A series of vessel members connected end-to-end form a **vessel**. Secondary walls are laid down after the primary wall and to the inside of the primary wall. Lignification of cell walls occurs when lignin is deposited in the walls. Pits are small openings that lack secondary walls. Tracheary elements are dead at maturity; they lose all cell contents, allowing for water to freely move through the tubes formed by these conducting cells.

In flowering plants, there is one type of sugar-conducting cell, the **sieve-tube member**. Although living at maturity, sieve-tube members lack nuclei when mature and are always associated with **companion cells**, which do have nuclei. Sieve-tube members are elongated cells with primary walls only, and when these cells are aligned end-to-end, they form **sieve** tubes. The sides of sieve-tube members are dotted with **sieve areas** and the end walls are covered in a **sieve plate**. The sieve areas and sieve plates contain small pores through which the cytoplasm is connected between adjacent cells. A carbohydrate, **callose**, plugs the pores of the sieve plates if the cell is disturbed, and prevents the contents from leaking out. Within the sieve-tube members, strands of **P-protein** are sometimes visible.

For this portion of the exercise, examine the prepared slides or prepare freehand sections of living plant material and make wet mounts, stained or unstained, to view living tissues and cells. Your instructor will demonstrate the proper method for making freehand sections. However, remember that a fresh single-edge razor blade gives the best results (and is less likely to cause injury). Two stains are employed in this exercise, Toluidine blue and phloroglucinol. Toluidine is an all-purpose stain and can be used throughout the course. Phloroglucinol specifically stains lignin red or reddish-pink, and is useful for distinguishing lignified sclerenchyma cells or tracheary elements from nonlignified cells.

MATERIALS

- prepared slide of *Pelargonium* leaf epidermis, whole mount
- two-week-old bean seedlings
- celery petiole
- pear fruit
- prepared slide of *Phaseolus vulgaris* stem, longitudinal section, vascular tissue
- prepared slide of *Cucurbita pepo* stem, longitudinal section, sieve tubes
- macerated oak wood in dropper bottle
- prepared slide of *Helianthus* stem, longitudinal section
- single-edge razor blades
- Toluidine blue stain in dropper bottle
- phloroglucinol-hydrochloric acid stain in dropper bottle
- compound microscope
- microscope slides
- coverslips
- dH$_2$O in dropper bottle

PROCEDURE

1. Mature Cell Types

   a. Epidermal cells

   First view the prepared slide of *Pelargonium* leaf epidermis on medium power of a com-

pound microscope, then change to high power. Note the shape of the epidermal cells and how they fit together.

b. Parenchyma cells

With a fresh single-edge razor blade, cut the stem of a young bean seedling about 1 inch below the apex. Invert the stem, and cut thin cross sections from it. Mount the sections in water on a microscope slide, add a coverslip, and view at low power of a compound microscope. If the sections are good, remove the slide from the microscope stage and stain the tissue with Toluidine blue. To stain the tissue, place drops of Toluidine blue on the slide next to one edge of the coverslip while holding paper toweling against the opposite side of the coverslip. This will draw the stain across the tissue. After staining, return the slide to the microscope stage and view the tissue at high power. Concentrate on cells in the center of the stem. These are parenchyma cells. Note their thin, primary cell walls and the cell contents.

c. Collenchyma cells

Using a sharp, single-edge razor blade, cut a very thin cross section through a celery petiole ("celery stalk") and mount the tissue in water on a microscope slide. View at low power of a compound microscope. If the section is good, change to high power and examine the celery tissue at high power. Clusters of collenchyma cells should be visible near the edges of the petiole just inside the epidermis. Note the uneven thickness of these cells and the pearl-like appearance of the primary walls. Parenchyma cells are also visible in the celery petiole. Staining the preparation with Toluidine will enhance the cellular structure.

d. Sclerenchyma cells

To view sclerenchyma cells, scrape tissue from a ripe pear fruit, discarding the "skin," and mount the tissue in phloroglucinol-hydrochloric acid stain on a microscope slide. (Take care with the stain, it can burn holes in clothing.) Locate the tissue with the low power of a compound microscope, then change to high power to view the cells. The sclerenchyma cells in pear fruit are sclereids, also called stone cells, and will stain red with the phloroglucinol stain. Sclereids give pears their gritty texture.

e. Conducting cells

(1) Use the *Cucurbita pepo* and *Phaseolus vulgaris* slides to view and compare conducting cells in longitudinal view (Figure 3-3).

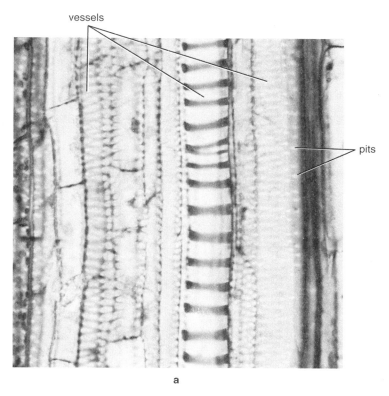

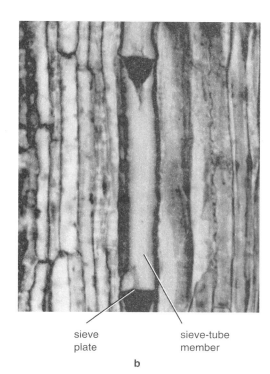

vessels

pits

sieve plate

sieve-tube member

a

b

Figure 3-3 Vascular tissue in longitudinal views. (**a**) Xylem. (**b**) Phloem.

Start with the lowest power objective lens on the compound microscope and progress to medium and higher powers to see details. Both slides have xylem and phloem tissues, although the *Cucurbita* slide is particularly good for viewing phloem tissue. Concentrate on finding sieve-tube members and vessel members. The sieve-tube members are very long and sieve plates should be visible in some of the cells. The vessel members are also elongated cells and some may reveal perforation plates. Look for pits along the side walls. In these prepared slides, the vessel members are stained red and the sieve-tube members are stained bluish-green.

(2) Obtain a dropper bottle containing macerated oak wood and make a wet mount of the macerated solution. Examine the preparation at high power and locate vessel members. The macerated cells are viewed in three-dimension and reveal perforation plates and pits in the side walls.

2. Mature Tissue Types in Plants with Primary Growth

   a. Epidermis

   Use the *Pelargonium* leaf prepared slide to study the epidermis tissue. You have examined the epidermal cells, now locate guard cells. Each stoma is formed of two guard cells with a pore between the cells. Note the distribution of the stomata across the leaf surface. Also, note the arrangement of epidermal cells around the paired guard cells, and look for epidermal outgrowths called trichomes.

   b. Ground tissue

   Use the prepared slide of *Helianthus* stem in cross section to view the location and distribution of ground tissue within a mature primary plant body. In this slide, lignified cells and nuclei stain red and cellulose stains green or bluish-green. The ground tissue is located in two regions of the *Helianthus* stem, the **cortex** and the **pith**. The cortex is the region immediately inside the epidermis but outside the ring of vascular bundles. The pith is the region in which the vascular bundles are embedded and the region in the center of the stem. The regions of pith tissue between vascular bundles are pith rays.

c. Vascular tissue

Continue using the *Helianthus* stem. Note the ring of vascular bundles. Vascular tissue in discrete strands that form a ring of bundles is the typical layout of vascular tissue within dicot stems. The red-stained tracheary elements are visible to the inside of each vascular bundle, and the green-stained phloem fills the outer portion of the bundle. Note the difference in appearance of the vascular tissue cell types in cross sectional view. Cell types other than tracheary elements and sieve-tube members may be apparent in the vascular bundles. Companion cells are identifiable as small cells adjacent to sieve tubes. In addition to vessel members and tracheids, xylem tissue contains parenchyma cells and fibers. Phloem tissue contains sieve-tube members, companion cells, parenchyma cells, and fibers. Parenchyma cells are identifiable by their cell contents. Fibers may not be present in this preparation, but if they are present, the cells will have a small diameter with relatively thick walls that stain red.

QUESTIONS FOR THOUGHT AND REVIEW

1. Name the cell type found within an apical meristem? What is the shape of these cells? _____

2. Are there any differences in the shapes and sizes of meristematic cells within shoot and root apical meristems? _____

3. List the three primary meristems and the mature tissues they produce. _____

4. Compare the overall shapes (length and width) of epidermal and parenchyma cells. _____

5. Compare the overall shapes of collenchyma and sclerenchyma cells. _____

6. List the conducting cell types of flowering plant tissue. Which conduct water? Which conduct carbohydrates? _____

7. List all cell types potentially present in xylem tissue. _____

In phloem tissue. _____

8. What mature cell types can be found in ground tissue? _____

9. What are the functions of parenchyma cells? Of collenchyma cells? Of sclerenchyma cells? _____

10. Describe one way in which the epidermis contributes to the survival of the plant body.

Name: _____

Section Number: _____

PRIMARY GROWTH: PLANT CELL AND TISSUE TYPES

1. What tissue gives plants the potential for eternal youth? _____

2. List the three primary meristems.

    a. _____

    b. _____

    c. _____

3. What are the two types of tracheary elements?

    a. _____

    b. _____

4. Sieve-tube members conduct _____

5. Parenchyma cells are the most common cell type in the mature tissue type called _____

    _____

6. Describe a function of collenchyma cells. _____

    _____

    _____

7. In plant cells with secondary walls, what is the location of the secondary wall relative to the primary wall?

    _____

    _____

8. In plant cells, what is a pit? _____

    _____

EXERCISE 4

ROOTS: PRIMARY AND SECONDARY GROWTH

OBJECTIVES

1. Differentiate between tap and fibrous root systems.

2. Identify the growth zones and the primary tissues of a root with primary growth.

3. Identify the cell types in a primary root when viewed in cross section and longitudinal section.

4. Describe the function of the endodermis in water and mineral uptake.

5. List two functions of the pericycle.

6. Identify a lateral root and an adventitous root from a root or stem cross section.

7. Describe the origin of lateral roots.

8. Understand the origin of the root vascular cambium and describe its development.

9. If given suitable specimens, distinguish between primary and secondary growth in roots.

10. Describe at least two root modifications.

TERMINOLOGY

adventitious root	primary root
apical meristem	primary xylem
Casparian strip	root cap
cell division zone	root hair
cork	root nodule
cork cambium	root primordium
cortex	secondary growth
elongation zone	secondary phloem
endodermis	secondary xylem
epidermis	stele
fibrous root system	suberin
growth zones	tap root system
lateral root	vascular cambium
maturation zone	vascular cylinder
pericycle	velamen
primary phloem	

PRIMARY AND SECONDARY GROWTH IN ROOTS

Roots and stems may appear superficially similar, but there are significant differences in their structure and function. Roots evolved to anchor the plant body in the soil, to absorb water and minerals from the soil, to conduct water and minerals from the region of absorption to the stem, and to store starch. Roots of herbaceous plants generally are limited to primary growth—growth that is the result of root **apical meristem** activity. However, woody plants have additional meristematic regions that increase their diameter. The structural origin and result of both primary and secondary root growth will be examined in this exercise.

PART A STRUCTURE OF THE PRIMARY ROOT

The first root produced by an embryonic plant is the radicle. The radicle emerges through the seed coat during seed germination and elongates into the **primary root**. **Lateral roots** develop from the primary root and from older lateral roots, eventually forming a root system. If the primary root thickens and is the dominant root within a root system, the root system is referred to as a **tap root system**. If the primary root remains similar in size to the numerous lateral roots, the system is a **fibrous root system**. **Adventitous roots** are common additions to fibrous root systems.

Young roots are ideally shaped to push through the soil as they grow, and their morphology and internal anatomy promote uptake of water and selective absorption of minerals from the soil. The root apical meristem produces new cells toward the apex of the root which mature into root cap cells and new cells away from the apex of the root which mature into the cells of the root axis. The root apical meristem is protected by the **root cap**, whose cells are sloughed off and continually replaced as the root grows through the soil. Behind the root apical meristem, cells elongate and begin differentiation, completing the process in the region of maturation.

MATERIALS

- plant with taproot system
- plant with fibrous root system
- two-day-old *Agrostis* seedling
- prepared slide of *Allium* root in longitudinal section
- water hyacinth plant with lateral root primordia
- microscope slide
- dissecting microscope
- compound microscope

PROCEDURE

1. Root Systems

   In the first exercise you examined tap and fibrous root systems. Examples of these living root systems are available in this exericse as a reference as you study the structure of roots in more detail. In the root system examples, locate the primary roots, lateral roots, and any adventitious roots present.

2. Growth Zones

   a. *Agrostis* seedling

      Gently place a two-day-old *Agrostis* seedling in a petri dish or on a microscope slide. To prevent the delicate structures from drying, keep the seedling moist, and, with the aid of a dissecting microscope, identify the **growth zones** of the young primary root. The internal tissues are visible because the root is transparent at this stage. Starting at the tip of the root, locate the root cap. Protected by the root cap is the root apical meristem and the **zone of cell division**. Behind the apical meristem is the **elongation zone**. Next is the **maturation zone** and the region of **root hairs**. Note the delicate nature and size of the root hairs. Root hairs are unicellular structures that mature from the protoderm. Recall that the protoderm is the primary meristem that develops into the **epidermis.**

   b. *Allium* root in longitudinal section

      Examine the prepared slide of the *Allium* root with a compound microscope, using the $10\times$ objective lens. Compare the growth zones visible in the prepared slide with those you observed in the living seedling. In Exercise 2 you used this slide to study the stages of mitosis in plant cells. Identify the growth zone of the dividing cells.

3. Lateral Roots

   Water hyacinth (*Eichhornia*) is useful for viewing the origin of lateral roots because the root tissues are transparent and reveal their inner structure. Observe the demonstration of the water hyacinth plant, and note the series of brown dots along the length of a root. The brown dots are **root primordia** that will form lateral roots. Root primordia originate deep within parent root tissue. As root primorida develop into lateral roots, they grow through the parent root tissue and break through the epidermis. Also note the long, brown root caps on the lateral roots that have emerged. Water hyacinth plants float in water (air-filled tissue in the leaf bases float the plant), and the root caps are not subjected to the sloughing action exerted by passage through the soil. Compare the size of lateral roots in water hyacinth with the *Agrostis* seedling root hairs. Lateral roots are multicellular structures; root hairs are single cells.

PART B   ROOT ANATOMY

The outer tissue of a primary plant root is the epidermis. Some cells of the epidermis are root hair cells. To the inside of the epidermis lies the **cortex** region, and the vascular tissue forms a solid cylinder in the center of the root. The **vascular cylinder** is also known as the **stele.** The epidermis matures from the protoderm, the cortex from ground meristem tissue, and the vascular cylinder from procambium. The cortex region in roots is mostly composed of parenchyma cells. At the innermost layer of the cortex, next to the vascular cylinder, is a cylinder of specialized tissue one cell wide. This is the **endodermis.** The endodermis tissue is specialized for controlling mineral accumulation in the root. Each cell of the endodermis contains **suberin** in a band encircling the transverse and radial walls of the cell. The bands of suberin are called **Casparian strips.** Water and minerals can freely pass through cellulose cell walls but cannot pass through suberized walls. The Casparian strips of the endodermal cells ensure that minerals must pass through a living cell where the plasma membrane can select what will pass into or out of the vascular cylinder. The outermost layer of cells in the vascular cylinder, next to the endodermis, is the **pericycle.** The pericycle retains its ability to divide longer than the surrounding cells, and it is the site of lateral root initiation. Within the vascular cylinder of primary roots, **primary xylem** and **primary phloem** tissue form alternating arms of tissue.

MATERIALS

- prepared slide of *Ranunculus* young root in cross section
- prepared slide of *Ranunculus* mature root in cross section
- prepared slide of *Ipomoea* young root in cross section
- prepared slide of *Smilax* mature root in cross section
- prepared slide of *Salix* root in cross section, showing lateral root origin
- prepared slide of *Lycopersicum* adventitious root in stem section
- compound microscope

PROCEDURE

1. Dicot Roots

   a. *Ranunculus* young root

   First examine the young root at low power of a compound microscope. Locate the epidermis, cortex, and vascular cylinder. The epidermis is one cell-layer thick, but the cortex is thick with many parenchyma cells that contain starch grains, reflecting the storage function of roots. Storage roots such as carrots and beets have extremely thick cortical regions. Note the degree of cellular maturation of the primary xylem tissue within the vascular cylinder. Mature conducting cells within the primary xylem are stained red in this slide. Immature vessel members remain green. Locate the pericycle and the endodermis and try to identify cell types within the primary phloem, particularly sieve-tube members and companion cells. Note that the cortex is composed of parenchyma cells.

   b. *Ranunculus* mature root

   Compare this slide with the previous slide. They are similar except that here the primary xylem tissue is mature (Figure 4-1). All of the conducting cells within the primary xylem are stained red, indicating complete lignification. Try to locate a Casparian strip along the endodermis. You may notice that some of the endodermal cells appear lignified. As roots age, lignin is deposited in the walls of endodermal cells and the cell walls appear thick.

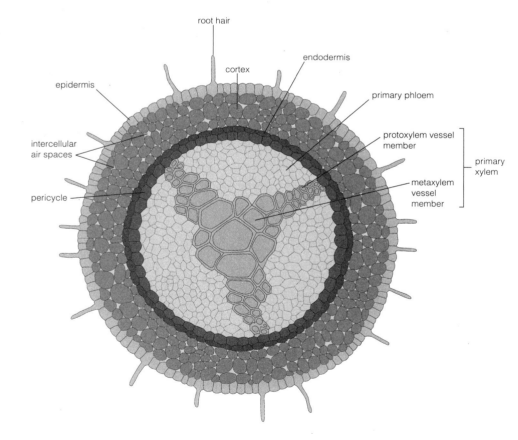

Figure 4-1 Sketch of root cross section through root hair zone, depicting tissues and cells of the dermal, ground, and vascular tissue systems.

root hair

cortex

endodermis

epidermis

primary phloem

intercellular air spaces

protoxylem vessel member

primary xylem

pericycle

metaxylem vessel member

c. *Ipomoea* young root

The Casparian strips in young *Ipomoea* roots are usually easier to see than in *Ranunculus* roots. Examine the slide on high power of a compound microscope and locate the endodermis. Scan the endodermis, looking for pink-staining Casparian strips. Also note that the number of primary xylem arms may differ from those visible in the *Ranunculus* root cross section.

2. Monocot Roots

Use a *Smilax* root cross section to study an example of a monocot root. Examine the slide with both low and high power objective lenses. Monocot roots, and some dicot roots, typically have a central region of parenchyma cells. This region looks similar to the pith region in dicot stems, but these parenchyma cells mature from procambium, not ground meristem. Consequently, the pith-like region of monocot roots is part of the vascular cylinder. The endodermis in mature *Smilax* roots is heavily lignified and easy to distinguish from surrounding parenchyma and pericycle cells. Compare the extent of the cortex region and the primary xylem with that of *Ranunculus* roots.

3. Origin of Lateral Roots

The *Salix* root slide has cross sections through the root zone where lateral roots arise from pericycle tissue. Use low power to locate a developing lateral root. Note the connections between the lateral root and the parent root vascular tissue. Also, note the plane in which the lateral root is sectioned. Lateral roots emerge at a right angle to the parent root, so the lateral root is seen in a longitudinal view in this preparation. Organs that arise deep within another organ are said to have an endogenous origin, and lateral roots are an excellent example.

4. Origin of Adventitious Roots

Adventitious roots arise from stems or leaves, not from other roots. Tomato plants commonly form roots from stem tissue. Examine the *Lycopersicum* stem section with adventitous roots. The tomato stem tissue does not have a pericycle, but adventitous roots are formed endogenously. Use high power of a compound microscope to look for the origin of an adventitous root within the tomato stem tissue. Note the connections between adventitous roots and the stem vascular tissue. Compare the origin of adventitious roots with that of lateral roots.

PART C SECONDARY GROWTH IN ROOTS

The root growth examined thus far has been primary growth. In primary growth, all cells originate from the apical meristem and its three primary meristems: the protoderm, the procambium, and the ground meristem. Once the primary tissues are mature, a plant cannot increase in girth. However, some plant stems and roots do increase in girth (for example, oak and pine trees). This type of growth is called **secondary growth**, and secondary growth is the result of cell divisions in a lateral meristem, the **vascular cambium**.

In roots, the vascular cambium originates from two groups of cells: the pericycle and procambial cells located between the primary xylem and primary phloem. In plants that undergo secondary growth, not all of the cells in the procambium differentiate into primary xylem and primary phloem. Some of the procambial cells retain their meristematic capacity. These cells are the residual procambium. In roots, the pericycle and the residual procambium together form the vascular cambium. The vascular cambium produces **secondary xylem** tissue to the inside and **secondary phloem** to the outside. As the vascular cambium grows and moves outward, increasing the plant's girth, the outer tissues are sloughed off. The epidermis, cortex, primary phloem, and old secondary phloem are eventually lost. A second lateral meristem, the **cork cambium**, forms and provides a protective outer layer of **cork** cells in secondary roots.

MATERIALS

- prepared slide of *Salix* root with secondary growth in cross section
- compound microscope

PROCEDURE

Examine the anatomy of the *Salix* root with secondary growth (Figure 4-2). The vascular cambium produces secondary xylem to the inside and secondary phloem to the outside. Woody roots look very similar to woody stems after many years of growth. Woody roots lack root hairs and no longer function in absorbing water. Woody roots also form a cork layer that replaces the epidermis. The formation of cork will be covered in the exercise on stems.

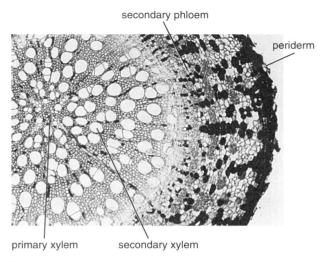

secondary phloem

periderm

primary xylem   secondary xylem

Figure 4-2 Cross section through a root with secondary growth. Note the vascular cambium and the positions of the secondary xylem and the secondary phloem relative to the vascular cambium.

PART D ROOT MODIFICATIONS

Several different types of root modifications, such as prop roots and storage roots, were studied in the first exericse. In this exercise, two additional modifictions will be examined. Some plants live in a mutualistic association with nitrogen-fixing bacteria. Only certain bacteria and cyanobacteria (both are prokaryotes) are capable of fixing atmospheric nitrogen. In legumes, nitrogen-fixing bacteria live in **root nodules**, which are large enough to be seen with the unaided eye. In the root nodules of leguminous plants, bacteria capture nitrogen from the air and use energy from the plant to convert the inorganic nitrogen to organic forms useful to the plant.

Another type of root modification is seen in the aerial roots of some orchids. Orchids typically grow as epiphytes attached to tree branches, but they do not obtain water or nutrients from the trees. These epiphytic orchids manufacture their own food through photosynthesis. In aerial roots of many epiphytic orchids, the epidermis is several cell layers thick, and in some orchids the aerial roots contain photosynthetic cells. The multiple epidermis of an orchid is called a **velamen**, and, in addition to protecting the cortex, it may absorb water. Examples of root nodules and orchid roots with a velamen are provided for study in this portion of the exercise.

MATERIALS

- plant with root nodes
- prepared slide of *Glycine max* root with attached root nodule
- prepared slide of orchid root with a velamen
- compound microscope

PROCEDURE

1. Root Nodules

   Examine the demonstration of plants with root nodules, and locate nodules along the plant roots. Then examine a prepared slide of *Glycine max* (soy bean) roots with root nodules. You will need to use the high power objective lens to see the details within an individual nodule. Note which cells of the roots are infected with the bacterial cells, and note the change in the anatomy of the infected cell.

2. Aerial Roots of Orchids

   Obtain a slide of an orchid root with a velamen, and examine the preparation with a compound microscope. Note the extensive multiple epidermis—the velamen—that surrounds the cortex and vascular cylinder. Orchids are monocots and the central cells of the stele are parenchyma cells that appear pithlike.

QUESTIONS FOR THOUGHT AND REVIEW

1. List several functions of the root cap._____

_____

_____

_____

2. What would happen to root hairs if they were produced in the region of elongation instead of the region of maturation? _____

_____

_____

3. How many cells make up a root hair? _____

_____

4. What would an endodermal cell look like in three dimensions? Where would the Casparian strip be located? _____

_____

_____

5. What tissue gives rise to lateral roots?

_____

6. What tissues give rise to the root vascular cambium? _____

_____

7. List all cell types found in roots with primary growth._____

_____

_____

8. Define adventitious root. _____

_____

_____

9. How does the anatomy of a monocot root differ from that of a dicot root? _____

_____

_____

10. Of what advantage are root nodules to the plant? To the nitrogen-fixing bacteria?_____

_____

_____

_____

_____

Name:_____

Section Number: _____

ROOTS: PRIMARY AND SECONDARY GROWTH

1. List the two types of primary root systems.

    a. _____

    b._____

2. Define adventitious root. _____
    _____

3. How does the shape of a young root promote root function? _____
    _____
    _____

4. How many cells make up a root hair? _____

5. As it grows, a root apical meristem produces tissue in two directions. Why is this true for root apical meristems and not for shoot apical meristems? _____
    _____
    _____

6. In a typical dicot root, what is the central region of parenchyma tissue called?_____
    _____

7. What is the function of the root endodermis?_____
    _____
    _____
    _____

8. Lateral roots arise endogenously from what specialized root cell layer?_____
    _____

EXERCISE 5

STEMS: PRIMARY AND SECONDARY GROWTH

OBJECTIVES

1. Distinguish a stem from a root by looking at the external morphology of a plant organ.

2. Identify the cell types and tissues visible within a dicot and a monocot stem cross section.

3. If given a stem section, determine if the organ is from a monocot or a dicot.

4. Point out the locations where the cambium will arise in cross sections of dicot stems that have not formed a vascular cambium, and name the tissues that give rise to the stem vascular cambium.

5. Identify all primary and secondary tissues and meristems, including both lateral meristems, in cross sections of dicot stems that have undergone secondary growth.

6. Describe how the vascular cambium forms both secondary phloem and secondary xylem.

7. Describe the origin and the components of the axial and ray systems of a stem that has undergone secondary growth.

8. Define and identify wood and bark.

9. If given a suitable specimen, identify the cork cambium and its derivatives.

TERMINOLOGY

annual ring	fascicular cambium
apical meristem	fiber
axial system	fusiform initial
axillary bud	ground meristem
bark	growth ring
bud scale scars	heartwood
bud scales	interfascicular cambium
bundle scars	late wood
companion cell	leaf primordium
cork	lenticel
cork cambium	parenchyma cell
cortex	periderm
early wood	phellem
epidermis	phelloderm

phellogen	sapwood
pith	secondary phloem
pith ray	secondary xylem
primary phloem	sieve-tube member
primary xylem	trichome
procambium	vascular bundle
protoderm	vascular cambium
radial system	vessel member
ray initial	wood

PRIMARY AND SECONDARY GROWTH IN STEMS

Stems form the above-ground portion of a plant axis and they produce leaves and lateral buds. The primary functions of stems are to support the leaves, to conduct water and minerals from the roots to the leaves, and to conduct sugars from regions of manufacture (such as photosynthetic leaves) or from storage areas (such as storage roots) to regions in the plant where sugars are needed. This direction of transport is referred to as movement of materials from source to sink. Stems also contribute to storage, to the production of new cells to maintain a healthy plant, and to the production of flowers for sexual reproduction. Most herbaceous stems are photosynthetic, and some are highly modified, such as corms and tubers, which you studied in Exercise 1.

As you saw in Exercise 1, the tips of stems are covered in tiny **leaf primordia** and end in a shoot **apical meristem**. Like root apical meristems, shoot apical meristems are potentially indeterminate tissues; as long as the plant remains vegetative the apical meristems are capable of dividing indefinitely. Stems reveal the modular nature of plants. Each repeating stem module consists of a length of stem, the internode, and a node with one or more leaves and **axillary buds** in the axils of leaves.

The shoot apical meristem only produces cells in an axial plane, but a lateral meristem, the vascular cambium, produces cells in two planes, axial and radial. Many dicots are woody plants, and their thick stems are the result of secondary growth produced by lateral meristems. Plants with woody growth actually have two lateral meristems, the **vascular cambium** and the **cork cambium**. The cork cambium contributes to the formation of bark. Primary and secondary growth from the vascular cambium and cork cambium will be examined in today's exercise.

PART A PRIMARY GROWTH IN STEMS

The anatomy of a typical herbaceous dicot stem reveals an organ that differentiated from the three primary meristems: **protoderm**, **procambium**, and **ground meristem**. In cross sections, a dicot stem with primary growth reveals a ring of discrete vascular bundles embedded in tissue differentiated from the ground meristem and surrounded by an **epidermis**. Inside the ring of **vascular bundles** is the **pith** region and between the vascular bundles and the epidermis is the **cortex** region. The areas of **parenchyma cells** between the vascular bundles are known as **pith rays**. Within each vascular bundle, **primary xylem** tissue is located toward the inside of the bundle and **primary phloem** tissue is located toward the outside. In between the primary xylem and primary phloem, in tissue that will undergo secondary growth, is a region of residual procambium. In this portion of the exercise, you will review stem morphology and examine stem anatomy in herbaceous plants.

MATERIALS

- living dicot plant
- young, potted *Hesperocnide tenella*
- prepared slide of *Coleus* stem tip in longitudinal section
- prepared slide of young *Medicago* stem in cross section
- dicot seedling
- prepared slide of *Zea mays* stem in cross section
- single-edge razor blades
- microscope slides
- coverslips
- dH$_2$O in dropper bottle
- dissecting microscope
- compound microscope

PROCEDURE

1. Stem Morphology

   a. Dicot stems

   Examine the shoot system of the dicot plant on demonstration and review the general stem morphology by identifying internodes, nodes,

leaves, axillary buds, and terminal buds. Keep these structures in mind as you examine the internal tissues of a dicot stem.

   b. Stem vasculature

   Examine the *Hesperocnide tenella* plant demonstration. *Hesperocnide* is a relative of the stinging nettle and can irritate the skin, so be careful about touching the plant. When young, the stems of *Hesperocnide* are clear and the vascular bundles are naturally red, making it easy to follow the pattern of vascular bundles along a length of stem. Look for vascular bundles within the stem and follow their path through the stem and into the leaves.

2. Anatomy of Primary Growth in Dicot Stems

   a. *Coleus* stem tip

   The slide with a longitudinal section through the shoot tip of a *Coleus* stem was examined in Exercise 3 and is used here as a review and to study additional details of the tissues and cells. First, examine the slide with a dissecting microscope, using the highest power. Locate nodes and internodes and the shoot apex. Note the red and green colors of the tissue. These are not natural colors; the tissue is stained with the standard botanical stains safranin and fast green, a combination that stains lignin, waxy materials, and DNA red, and stains other parts of cells and tissues green. Move the slide to a compound microscope and, using the intermediate and high power objective lenses, examine the details of the tissues. Note the fragments of disconnected tissue that occur above the shoot tip. These are parts of young leaves. Locate the shoot apical meristem and the primary meristems, and look for maturing vascular tissue in the stem and older leaves. Compare parenchyma cells in the pith and cortex regions. Look for **trichomes** in mature regions of the epidermis.

   b. *Medicago* stem cross section

   Get a cross-section slide of a *Medicago* young stem and use the 10× objective lens of a compound microscope to initially examine the tissues. Locate the ring of vascular bundles and note the variation in bundle size (Figure 5-1). Focus on a large vascular bundle and examine its cellular structure at high magnification. Distinguish between primary xylem and primary phloem. Within the primary xylem, locate tracheary elements and parenchyma cells. Within the primary phloem, distinguish between **sieve-tube members**, **companion**

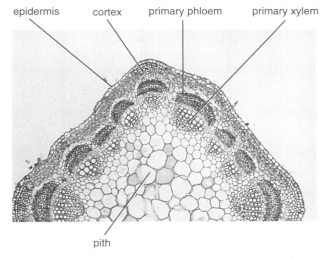

epidermis   cortex   primary phloem   primary xylem

pith

Figure 5-1 *Medicago* stem cross section. A young stem with discrete vascular bundles.

cells, parenchyma cells, and **fibers**. Examine the epidermis and look for trichomes and guard cells.

c. Seedling stem freehand section

With a fresh single-edge razor blade, cut a stem cross section from the youngest part of the stem of a dicot seedling. Your instructor will demonstrate the technique. Do not try to get slices of uniform thickness like slices of bread. Thin, uneven slices will serve. Starting halfway across the width of the specimen and cutting at a shallow angle can produce useful thin sections. Cuts will be easier to make if the tissue and the razor blade are kept wet. Place good sections in a drop of water or in a drop of diluted Toluidine blue stain on a microscope slide. Add a coverslip. If the coverslip is tilted, the sections are too thick. Start over. To avoid getting stain on the objective lenses, before placing the preparation on the microscope stage, blot excess stain from the slide with a paper towel. Look for vascular bundles in your sectioned material and compare it with the prepared slide.

3. Anatomy of Primary Growth in a Monocot

Obtain a prepared slide with a stem cross section of the monocot *Zea mays* (corn) and examine it with both the 10× and 40× objectives. Look for the scattered pattern of vascular bundles within the ground tissue (the terms cortex and pith are not used with monocots). Notice the orientation of the bundles: is xylem always oriented toward the center of the stem? Each bundle has a large air space in addition to the two large vessel members and many smaller vessel members. The air space arose when expansion of the bundle tore apart an earlier-formed vessel member. In some cases you may see remnants of the torn vessel member within the air space. Moving to the green-staining phloem region, you can easily pick out the sieve-tube members and companion cells; the two cell types occur side by side, with large sieve-tube members packed next to small companion cells in the corners between sieve-tube members.

PART B SECONDARY GROWTH: THE VASCULAR CAMBIUM IN STEMS

Many dicots are woody plants, and their thick stems are the result of secondary growth produced by lateral meristems. Specifically, **wood** is **secondary xylem**. The shoot apical meristem only produces cells in an axial plane, but the vascular cambium produces cells in two planes, axial and radial. The tissues produced in an axial plane form the **axial system** and tissues in the radial plane form the **radial system**. The addition of radial growth leads to an increase in the girth of stems. The residual procambium of vascular bundles and some of the parenchyma cells between the vascular bundles together form the vascular cambium, which is a lateral meristem. The vascular cambium forms a ring of tissue one to two cells in width, and it produces secondary xylem tissue to the inside and **secondary phloem** tissue to the outside.

Meristematic cells are sometimes called initial cells, and there are two types within the vascular cambium meristem: **fusiform initials** and **ray initials**. The two types of initials give rise to the axial and radial systems of secondary tissues. Fusiform means spindle-shaped, and fusiform initials give rise to the elongated cells of the axial system. The ray initials develop the radial (or ray) system. Within the axial system of secondary xylem, fusiform initials produce vessel members, tracheids, fibers, and axial parenchyma cells, and within the secondary phloem, fusiform initials produce sieve-tube members, companion cells, fibers, and axial parenchyma cells. The radial system's ray initials produce ray parenchyma cells in the secondary xylem and secondary phloem of flowering plants. Each vascular cambial initial, whether a fusiform or ray initial, can divide to add cells to the inside of the ring of vascular cambium or to the outisde of the vascular cambium. This unusual property leads to the development of massive amounts of secondary xylem on the inside of stems and smaller amounts of relatively short-lived secondary phloem to the outside. Due to the pattern of the vascular cambium's growth, all cells to the outside of it are eventually destroyed by this outwardly advancing lateral meristem.

The complex system of secondary growth shows its greatest regularity in the secondary xylem, or wood. The familiar **growth rings** visible in cross sections of woody stems are the result of differences in cells produced earlier or later in the growing season of temperate plants. The growth rings are actually **annual rings**, and each ring is the result of one year's growth. The width of the ring reflects differences in growing conditions from year to year. Larger diameter cells are produced earlier in the growing season, this is **early** or **spring wood**, and **late** or **summer wood** follows with cells of narrow diameters. A band of early and late wood forms one growth ring. The cells of very old wood no longer function in transporting water and minerals and may become filled with substances that turn the wood dark, this nonconducting wood is **heartwood**. **Sapwood** is younger secondary xylem that is still conducting water and minerals, or sap. In this exercise, we will examine the origin and development of the vascular cambium, secondary xylem, and secondary phloem. We will examine the wood of an oak tree in detail.

MATERIALS

- woody stems
- prepared slide of *Medicago* stem with early vascular cambium in cross section
- prepared slide of *Medicago* mature stem in cross section
- prepared slide of *Robinia* vascular cambium in tangential section
- prepared slide of two-year-old *Tilia* stem in cross section
- oak block demonstrating transverse, radial, and tangential cuts
- model of wood in three planes
- prepared slide of *Quercus* wood in transverse, radial, and tangential sections
- variety of cut pieces of wood
- compound microscope

PROCEDURE

1. External Features of Woody Stems

   The external morphology of young woody branches may have an appearance that reflects the perennial nature of the plant. The terminal buds and lateral buds of overwintering woody branches are enclosed within modified leaves, the **bud scales**. When the terminal buds resume growth in the spring, the terminal bud scales drop away and leave characteristic terminal **bud scale scars**. Examine the woody stem specimens on display and locate terminal buds, terminal bud scales, and terminal bud scale scars along older portions of the stem. The series of terminal bud scale scars can be used to determine the age of the specific stem. When a leaf is dropped from a stem (a process called abscission) the vascular bundles may leave **bundle scars** on the stem.

2. Initiation of the Vascular Cambium in Dicot Stems

   Examine the *Medicago* stem cross section labeled as having early vascular cambium. Focus on two side-by-side vascular bundles and examine the region with the 40× objective. The residual procambium between the primary xylem and primary phloem within the bundle has just started differentiation into a portion of the vascular cambium called **fascicular cambium**. (Fascicle is another term for a bundle.) Parenchyma cells between the vascular bundles and aligned with the fascicular cambium are beginning to dedifferentiate and form a portion of the vascular cambium called the **interfascicular cambium** (between bundles). Look for signs of recent cell divisions within the region of the developing vascular cambium. The fascicular and interfascicular cambia eventually unite to form a complete cylinder of vascular tissue one or two cells in width. This is the complete vascular cambium; it will produce secondary vascular tissues.

3. Anatomy of the Vascular Cambium

   a. *Medicago* mature stem with early vascular cambium

      Examine the *Medicago* mature stem cross section with the 10× and then the 40× objective lenses of a compound microscope (Figure 5-2).

secondary phloem   vascular cambium

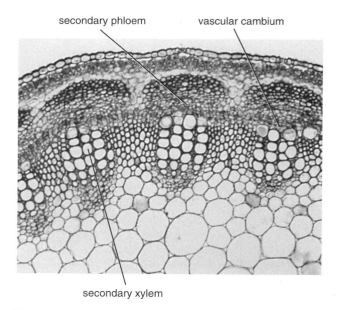

secondary xylem

Figure 5-2 *Medicago* stem cross section. A mature stem with secondary growth.

Locate the complete vascular cambium and compare it with the previous slide. Note the extensive secondary xylem, but also note that the primary vascular bundles can still be discerned. The cells of the secondary xylem form in single files. Look at the phloem and try to distinguish secondary from primary phloem. Note the numerous phloem fiber cells capping the regions of primary phloem.

b. *Robinia* vascular cambium

The *Robinia* slide of vascular cambium is a type of longitudinal section. The slide reveals the axial and radial systems and their initial cells, fusiform and ray initials. Portions of the slide include initials that are beginning to mature into conducting cells. A different stain was used with this tissue. Sieve plates are stained blue and are distinctive. In this preparation, the length and width of cells in the axial system can be determined. In addition, the size of rays and number of parenchyma cells in each ray are easy to see. First examine the slide with the 10× objective and then switch to higher power to examine details of individual fusiform and ray initial cells. Attempt to distinguish between initial cells and cells beginning to differentiate into cells of the secondary xylem or secondary phloem.

c. A two-year-old *Tilia* stem

The *Tilia* two-year-old stem reveals extensive regions of secondary xylem and secondary phloem. Examine the slide with low then high power objective lenses. Note the extent of the secondary xylem and try to determine the stem's age by counting growth rings. *Tilia* has especially interesting secondary phloem. Examine the phloem rays and the layered tissues between the rays. Deduce which areas have fibers, parenchyma cells, or conducting tissue. The phloem rays in *Tilia* are called dilated rays. The parenchyma cells continue to divide and help compensate for the increase in girth of the stem. Eventually, these cells are sloughed as the vascular cambium advances.

4. Anatomy of Secondary Xylem in Oak

a. Oak block

The two systems of cells, the axial system and the radial system, can be studied in a block of wood from a small-diameter oak tree. Different views of the axial and radial systems are produced when the organ is sectioned in different planes. Examine the oak (*Quercus*) wood block (and the available model) and try to visualize the cells within the two systems.

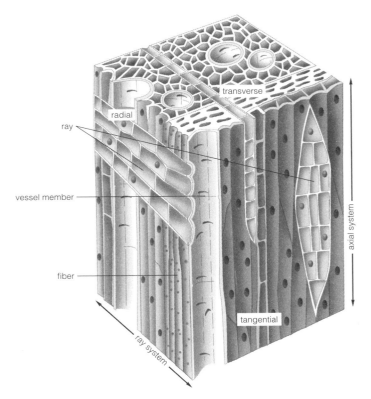

Figure 5-3 Three-dimensional diagram of red oak showing cells as they appear in cross section, radial section, and tangential section.

b. Transverse, radial, and tangential sections of oak wood

Obtain a slide containing *Quercus* wood in cross, tangential, and radial sections (Figure 5-3, page 35). A cross section is made at right angles to the axial system. Tangential and radial sections are made parallel to the axial system. Start with the cross section, it is easy to identify. Examine the slide with higher power of a compound microscope. Some of the vessels are huge; others are relatively small. Masses of slender fiber cells are also present, giving oak wood its hardness. Look for rays, made of parenchyma cells. Continue the study by looking at the two longitudinal sections; radial and tangential. The radial section is cut parallel to a radius and the tangential section is cut along a tangent through a radius. Radial and tangential sections can be distinguished by looking at the cells of the rays: a radial section will show rays from a side view and the ray parenchyma cells will look like stacks of bricks; tangential sections show rays in end view. Some rays are narrow, only a few cells wide; others are large. Use logic and imagination to decide which view you are seeing as you compare sections.

5. Wood Diversity

Wood is composed of secondary xylem. The types of cells and the ratio of the cell types in secondary xylem varies among different species. Because of the differences in cell types and ratios, wood can be hard or relatively soft. Examine the wood specimens on display and offer a hypothesis for the differences in their hardness.

PART C SECONDARY GROWTH: THE CORK CAMBIUM IN STEMS

Because of the advancing front of vascular cambium, all tissues to the outside of the epidermis and cortex are eventually sloughed from stems with secondary growth. However, the stem still requires a protective barrier from the environment. In stems with secondary growth, the epidermis is replaced by **cork**. The lay person refers to this outer covering as **bark**, but bark is a technical botanical term that encompasses all tissues to the outside of the vascular cambium in woody plants. Therefore, bark includes secondary and primary phloem, any remaining cortex, and the cork and associated layers. The cork tissues are produced by an additional lateral meristem, the cork cambium. The first cork cambium, and there may be many, usually originates from parenchyma cells within the outer cor-

tex. A ring of parenchyma cells dedifferentiates, becoming meristematic, and form the initial cells of the cork cambium lateral meristem. If additional cork cambia are produced in progressively older stems, they are produced from deeper secondary phloem parenchyma cells. All cork initials are similar in shape (they are isodiametric); there are no axial or radial systems produced from the cork cambium. The cork cambium initials produce cork cells to the outside. Mature cork cells are heavily suberized or lignified. To the inside of the cork cambium, cork initials produce parenchyma cells. (Sometimes the cork cambia is called **phellogen**, the cork cells **phellem**, and the parenchyma cells **phelloderm**.) Collectively, cork cells, cork cambia, and parenchyma produced by cork cambia are called the **periderm**.

Stomata provide for gas exchange in the epidermis of young, herbaceous tissue. In periderm, gas exchange occurs in **lenticels**. Lenticels are regions of loosely packed parenchyma cells and are particularly visible on smooth-bark trees as small, lens-shaped slits in the cork.

The cork of woody plants is variable and can be used to help identify plant species. For example, the cork of cork oaks, from which bottle corks are made, is uniform for extensive depths. Some cork layers peel from the tree in large sheets of thin cork tissue and others split to form deep fissures in the outer layers.

MATERIALS

- prepared slide of *Sambucus* young stem with early cork in cross section
- prepared slide of *Sambucus* with mature bark and lenticels in cross section
- smooth cork stem with prominent lenticels
- variety of bark pieces displaying different forms of cork
- dissecting microscope
- compound microscope

PROCEDURE

1. Anatomy of the Stem Periderm

   a. Origin of cork cambium

   The *Sambucus* young stem with early cork formation offers a clear view of the cork cambium and rows of early cork cells. Examine the slide with the high power of a compound microscope. Locate the cork cambium and follow the files of cork cells. This is a young stem, so the epidermis may still be present and much of the cortex will be present. Note the region in which the cork cambium arose.

b. Lenticels

Examine the display of woody stems with lenticels, particularly noting the shape and location of lenticels throughout the specimen. Then examine the prepared slide of *Sambucus* stem with mature bark and lenticels. Use low power of a compound microscope. Lenticels will appear as torn regions along the edge of the stem. Note the parenchyma cells within a lenticel. Also, note the extensive cork in this preparation.

2. Bark Diversity

Examine the display of bark diversity. Feel each specimen and note which are smooth and which are rough. Try to determine if a particular sample has a single periderm or multiple periderms. You may wish to examine some of the specimens with a dissecting microscope.

QUESTIONS FOR THOUGHT AND REVIEW

1. What external features distinguish a stem from a root? _____

_____

2. What cell type would you expect to find in the center of a dicot stem? From what primary meristem would these cells mature? _____

_____

3. What anatomical features distinguish dicot from monocot primary stems? _____

_____

_____

4. Do dicot stem vascular bundles follow a pattern, or are they randomly spaced within the stem?

_____

5. What is the function of the vascular cambium?

_____

_____

_____

6. How can you determine the age of a woody stem?

_____

_____

7. How can you distinguish vessels from parenchyma cells in the three different types of wood sections?_____

_____

_____

8. What is the difference between a pith ray and a wood ray? _____

_____

_____

9. What distinguishes wood from bark? _____

_____

_____

_____

10. What structure allows for gas exchange in a woody stem?_____

_____

_____

Name:_____

Section Number: _____

STEMS: PRIMARY AND SECONDARY GROWTH

1. An axillary bud is found in the axil of what organ? _____

2. List three important functions of a typical stem.

   a. _____

   b. _____

   c. _____

3. Unlike lateral roots, which emerge from deep within an older root, lateral stems are produced from _____

_____

4. Primary growth results from the activity of what tissues? _____

_____

5. When viewing a cross section of a typical herbaceous stem, you observe a distinctive ring of tissue in bundles. What are these bundles of tissue called, and what is their major function? _____

_____

_____

6. Secondary growth results from the activity of what two lateral meristems? _____

   a. _____

   b. _____

7. Contrast cork and bark. _____

_____

_____

8. What is a growth ring? _____

_____

_____

EXERCISE 6

LEAVES

OBJECTIVES

1. Describe the morphology of simple and compound leaves.

2. Describe the difference between netted and parallel venation.

3. Recognize and describe the cells and tissues of a typical mesophytic leaf, and explain the functions of the tissues and cells.

4. Describe the adaptive features of a hydrophytic leaf and a xerophytic leaf and explain the benefits and disadvantages of each feature.

5. Recognize Kranz anatomy and bulliform cells. Describe the functions of these structures and cells.

TERMINOLOGY

abscission
blade
bulliform cell
bundle sheath
compound leaf
cuticle
guard cell
hydrophyte
Kranz anatomy
lamina
leaf primordium
leaflet
mesophyte
midrib
netted venation
palisade mesophyll

palmately compound
paradermal section
parallel venation
petiole
petiolule
pinnately compound
rachis
sessile
simple leaf
spongy mesophyll
stoma
stomatal crypt
transpiration
vein
venation
xerophyte

INTRODUCTION TO THE LEAF OF FLOWERING PLANTS

In most flowering plants, the leaf is the chief organ of photosynthesis, and it typically comes in two parts: the leaf stalk, called the **petiole**, and the expanded, flattened portion called the **blade** or **lamina**. Leaves, unlike stems and roots, are not indeterminate structures. They reach a mature size and shape and then cease growing. Despite the restraints of determinate growth, plants display enormous varieties of leaf morphology and anatomy among different species and sometimes on the same plant. Accordingly, the external and internal structure of a leaf often reveals the habitat of the plant that produced it. This is because leaves exhibit adaptations for surviving in environments ranging from hot and dry to submersion under water.

We can classify plants into groups based on three general environmental conditions. **Xerophytes** are plants that grow in arid environments, **mesophytes** grow where there is a reliable soil water source and a relatively high humidity, and **hydrophytes** grow partially or wholly submersed in water. During today's exercise, we will explore examples of variation in leaf morphology and anatomy and relate them to environmental conditions.

PART A LEAF MORPHOLOGY

As you saw in the first exercise, dicot leaves are either simple or compound. A typical **simple leaf** has a petiole and a blade, and the blade, although it may be deeply lobed, is never divided into distinct parts, or leaflets. In contrast, the blades of **compound leaves** are divided into **leaflets**. Compound leaves are either **pinnately compound** or **palmately compound**. In a palmately compound leaf, all leaflets diverge from the tip of the petiole, much like fingers from the palm of a hand. In a pinnately compound leaf, the extension of the petiole to which leaflets are attached is the **rachis**, and each leaflet generally has a short stalk called a petiolule. Leaflets extend from the rachis like the pinnae of bird feathers. (Leaf blades that directly attach to the stem and lack a petiole are said to be **sessile**.) It is actually easier to distinguish between leaves and leaflets than it may first appear. Axillary buds are present in the axils of leaves, but not leaflets; and, on a given plant, leaves may be oriented in different planes, but leaflets of a leaf always lie in the same plane.

In addition to simple, pinnately compound, and palmately compound morphology, leaves are distinguished by the distribution of the veins within a given leaf. The distribution of **veins** within a leaf or leaflet is called **venation**. The two general categories of leaf venation are **netted** and **parallel**. Although not always so, netted venation is typical of dicots and parallel venation is typical of monocots. Two subtypes of netted venation are pinnately or palmately netted.

MATERIALS

- examples of simple and compound leaves
- examples of dicot and monocot leaves
- mounted, cleared leaf or leaf skeleton
- dissecting microscope

PROCEDURE

1. Leaf Shape

   Review the examples of leaves representing various morphological shapes. Be certain you can distinguish between simple and compound leaves and between dicot and monocot leaves.

2. Leaf Venation

   a. Living leaves

   Again, look at the leaves on display, but this time examine their venation. Decide if a given leaf or leaflet has netted or parallel venation, and if the venation is netted if it is pinnately netted or palmately netted.

   b. Cleared leaf

   Cleared leaves (or leaf skeletons can be used) are useful in studying the distribution of veins within a leaf. Observe the demonstration of a cleared leaf, and look for veins and vein endings. Notice the many levels of vein branching, and the relationship between variation in vein size and distribution of veins. You may want to use a dissecting microscope to study details. Note if any large areas of tissue are not served by a vein. Determine if the specimen displayed has netted or parallel venation.

PART B STRUCTURE OF A TYPICAL MESOPHYTIC LEAF

Leaves originate from cells near the surface of the growing shoot tip. This position of origin is exogenous and very different from the endogenous origin of lateral roots. As a leaf matures, the cells expand and differentiate, and a typical mature mesophytic leaf is a flattened blade with a petiole. The internal structure of the blade reveals an internal region of ground tissue in which vascular tissue is embedded and all covered with an epidermis. The surface of the epidermis is coated with a waxy layer, the **cuticle**, that is relatively impervious to water and gases. Pores in the epidermis, **stomata**, allow for gas exchange. Carbon dioxide must be taken in for photosynthesis and when stomata are open, water vapor can escape. The loss of water vapor from the surface of a leaf is **transpiration**. The ground tissue, or mesophyll, consits of two regions: palisade parenchyma tissue and spongy parenchyma tissue. The upright, **palisade mesophyll** is adjacent to the upper epidermis; the **spongy mesophyll** is adjacent to the lower epidermis. Most leaves are short-lived, although some live for many years, but eventually most leaves age and drop from the stem. This process is called **abscission**. Plants that lose all of their leaves during the fall are winter deciduous, those that lose their leaves during the hot, dry part of the year in arid regions are drought deciduous, and those that keep leaves for years and lose a few at a time are evergreen. During this portion of the exercise, we will follow the development of a leaf from a protuberance on the edge of a shoot apex to the mature form to the stage of abscission.

MATERIALS

- prepared slide of *Syringa* stem tip in longitudinal section
- prepared slide of *Syringa* leaf in cross section
- prepared slide of *Syringa* leaf in paradermal section
- prepared slide of *Populus* leaf abscission in longitudinal section
- compound microscope

PROCEDURE

1. Ontogeny

   Examine a prepared slide of *Syringa* (lilac) stem tip in a longitudinal view. With low power of a compound microscope, locate the shoot apex, then change to high power to locate a leaf primordium. Divisions beneath the protoderm of the shoot apex lead to the formation of a bulge that will increase in cell number and enlarge to form a leaf primordium. Once the primordium is established, intercalary growth creates the blade and petiole. Examine the simple anatomy of the primordium, then locate an older leaf further down the shoot tip. Note if any procambium or vascular tissue is visible within the center of the older leaf. Increasingly older leaves will reveal greater differentiation of cells.

2. Anatomy

Cross section and paradermal sections of *Syringa* will reveal a different view of leaves that have already matured. Combined, the views help in attempting to visualize the three-dimensional shapes of cells and tissues within a leaf.

a. Cross section of a lilac leaf

Obtain a *Syringa* leaf cross section slide and observe the tissues and cells with a compound microscope (Figure 6-1). Start with the lowest power and increase power to see more detail. At low power, scan the slide to see the overall structure of the leaf's anatomy. Note the upper epidermis, palisade parenchyma cells, veins, spongy parenchyma cells, and lower epidermis. Focus on the upper epidermis and note the cuticle layer on the outer surface of this tissue. Compare the shapes and distribution of cells within the mesophyll and note the location of veins relative to the palisade and spongy tissues. Also, note the large vein in the center of the leaf. This is the leaf **midrib**. In

the lower epidermis look for stomata. Each stoma has two **guard cells** that flank a pore. The guard cells control the size of the opening in the pore, from closed to wide open. As you study the cross section, look for features that would help the leaf to resist collapse, mechanical damage, and water loss, as well as features that would aid in capturing and moving carbon dioxide, venting off heat, and passing materials to and from the photosynthetic cells.

b. Paradermal section of a lilac leaf

The **paradermal section** of lilac reveals a surface view of each tissue layer. Using low power, locate the lower epidermis. The cells will be shaped like tight-fitting pieces of a jigsaw puzzle and stomata will be visible. Change to higher power and view the tissue and cells. The next tissue region to the inside is the spongy mesophyll. The cells are irregular in shape but have many and sometimes large intercellular spaces between each parenchyma cell. Between the spongy tissue and the upper epidermis is the palisade mesophyll. These cells are more regular in outline and, although intercellular spaces are abundant, the spaces are less extensive in size than in the spongy mesophyll. Throughout the mesophyll, leaf veins are visible. Note the differences in the sizes of the veins. Some are large, others just one or two cells. Most veins are surrounded by a bundle of parenchyma cells called a **bundle sheath**.

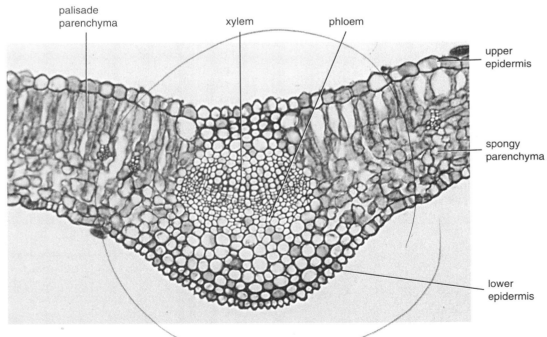

Figure 6-1 *Syringa* leaf cross section. Note the upper and lower epidermis, the palisade and the spongy parenchyma, and the large midrib vein.

3. Leaf Abscission

Prior to leaf drop, structural changes occur to weaken the connection of leaf and stem. Obtain a slide of *Populus* leaf abscission to observe the changes in a leaf close to the stage of leaf drop. Near the base of the leaf petiole a two-layered abscission zone is visible. One layer is the separation layer and the other is the protective layer. Within the separation layer, short cells with thin, degenerating cell walls make the region structurally weak. In the protective layer, closer to the stem, the cells become suberized. The suberized cells seal the region after leaf fall. Locate the two regions of the abscission zone and examine the anatomy of these tissues and cells. The protective layer leaves a visible leaf scar on the stem.

PART C VARIATION IN THE STRUCTURE OF LEAVES

Variation in leaf morphology can reflect the variation in plant habitat. For example, plants growing submersed or floating in water may have a different morphology and internal structure compared with plants growing in arid regions. Other plants evolved characteristics that reflect a difference in the biochemical processes of photosynthesis. Many woody plants produce individual leaves with different anatomy depending upon whether the leaf is growing under low light intensity or high light intensity; and, both "shade" and "sun" leaves are on the same individual.

Hydrophytes are plants adapted for growth in water. Because an aquatic habitat is more uniform than arid terrestrial environments, there is less diversity of structural adaptations than within the xerophytes. However, typical hydrophytic adaptations include stomata on the upper surface of floating leaves, thin epidermal cell walls of cellulose with little or no cuticle, and large intercellular spaces filled with gases.

Xerophytes are plants that grow in climates with extremely low rainfall during all or part of the year or where the climate is cold and abundant water is unavailable for much of the year. Not all plants growing in arid regions exhibit xerophytic adaptations. For example, plants in arid climates may only grow during the rainy season. But some arid-region plants evolved adaptations to protect against excessive water loss throughout the year. Three common traits of xerophytes are a thick cuticle, masses of sclerenchyma tissue, and sunken stomata. Many grass leaves exhibit a xeric adaptation where the leaves roll up to reduce water loss. These leaves control unrolling and rolling with specialized motor cells called **bulliform cells**. In this portion of the exercise, we will examine the leaf specimens exhibiting variation in leaf morphology and internal anatomy.

MATERIALS

- prepared slide of *Nymphaea* leaf in cross section
- prepared slide of *Yucca* leaf in cross section
- prepared slide of *Poa* leaf in cross section
- prepared slide of *Saccharum* leaf in cross section
- prepared slide of *Acer* sun and shade leaves in cross section
- compound microscope

PROCEDURE

1. Leaf from a Hydrophytic Environment

Examine the *Nymphaea* (water lily) leaf cross section with a compound microscope. Start with low power to view the overall anatomical structure. The leaves of water lilies float on the surface of the water. Scan the upper epidermis and compare it with the lower epidermis. All of the stomata are in the upper epidermis and the cuticle is extremely thin. Note that the vascular tissue is not as extensive as in the lilac leaf and that there are large intercellular air spaces between the spongy parenchyma cells of the *Nymphaea* leaf. The large air spaces provide buoyancy for the floating leaf. Examine the palisade parenchyma tissue, it is several cell layers thick. There are occasional large, branched sclereids within the mesophyll tissue.

2. Leaf from a Xerophytic Environment

With a compound microscope examine a prepared slide of the xerophytic *Yucca* leaf in cross section. Note the thickness of the cuticle and of the outer epidermal cell walls, the sunken stomata in **stomatal crypts**, the reduced blade area, the bundles of fibers, and the extra water-storing tissue. (Sunken stomata are below the rest of the leaf surface and stomatal crypts are inward folds in the leaf surface.)

3. Grass Leaves

Grass leaves have parallel venation and a variety of epidermal cells including longitudinal rows of specialized cells involved in the rolling and unrolling of the leaf. This is a xeric adaptation that decreases water loss. There are differences in the anatomy of the leaves of C_3 and C_4 grasses. (The C_3 photosynthetic pathway is the standard photosynthetic pathway; the C_4 pathway evolved in some plants that grow in hot environments. During hot conditions, C_4 photosynthesis is more efficient.) In grasses with the C_4 pathway of photosynthesis the veins are closer together and mesophyll cells form a wreath of cells around the veins. In this portion of the exercise, we will compare the C_3 leaf characteristics of a grass leaf with bulliform cells with a C_4 grass leaf.

a. Bulliform cells

Examine the *Poa* grass leaf cross section and locate the greatly enlarged epidermal cells that function as motor cells, or bulliform cells. During times of drought, when excessive water is lost from the plant, these cells lose turgor and the leaf rolls. The *Poa* plant is a C_3 plant. Note the number and arrangement of mesophyll cells between and around the veins.

b. Kranz anatomy

Sugarcane (*Saccharum*) is a C_4 plant. Examine the prepared slide of a sugarcane leaf cross section. Count the number of mesophyll cells between veins and compare this with that seen in the *Poa* grass leaf. Note the radial arrangement of mesophyll cells around each vein. The wreathlike arrangement of mesophyll cells is called **Kranz anatomy** and is typical of many C_4 plants.

4. Sun and Shade Leaves

Environmental factors such as light can have a major effect on the development of leaves on the same plant. Trees often grow sun and shade leaves. Sun leaves are exposed to higher light intensity and are much thicker than shade leaves. The thicker leaf is due to more palisade parenchyma tissue. Shade leaves develop under lower light intensities and the leaves are much thinner. Examine the prepared slide showing both sun and shade leaves of *Acer* and note the differenes in the development of the mesophyll tissue and the comparative thickness of the leaves.

QUESTIONS FOR THOUGHT AND REVIEW

1. How does the upper epidermis differ from the lower epidermis in a typical mesophytic leaf? How might the differences relate to the function of the leaf? _____

_____

_____

2. What is a function of intercellular spaces within the mesophyll of a leaf? _____

_____

_____

3. Do the epidermal cells have chloroplasts? _____

4. In what tissue within a leaf is the greatest concentration of chloroplasts? _____

_____

5. What is the function of stomata?_____

_____

_____

6. How do gases enter the leaf of a submersed plant? _____

_____

_____

7. Why would there be less vascular tissue in a hydrophyte leaf than in a mesophyte leaf? _____

_____

_____

8. How do stomatal crypts and sunken stomata contribute to leaf water conservation? _____

_____

_____

9. How do bulliform cells contribute to water conservation? _____

_____

_____

10. Which is thicker, a sun leaf or a shade leaf?

_____

Why? _____

_____

_____

Name: _____

Section Number: _____

LEAVES

1. Describe how to distinguish a simple from a compound leaf. _____
_____
_____

2. What is a petiolule? _____
_____

3. Can leaf venation be used to distinguish dicot leaves from monocot leaves? _____
Explain. _____
_____
_____

4. What specialized name is given to parenchyma tissue within a leaf? _____

5. What is the primary function of leaf stomata? _____
_____

6. How do leaves obtain water? _____
_____
_____

7. List two anatomical features of a leaf specialized for living in a xeric environment.
    a. _____
    b. _____

8. What feature would you look for in a leaf cross section to determine if the leaf is from a C_3 or a C_4 plant?
_____
_____

PHOTOSYNTHESIS AND RESPIRATION

OBJECTIVES

1. Understand what is meant by visible light.

2. Describe what is happening to electrons during fluorescence of a pigment extract.

3. Name the fat-soluble photosynthetic pigments in plant chloroplasts.

4. Describe the main functions of photosynthesis and respiration.

5. Describe the significance to the life of a plant if the seed and seedling are kept in the dark.

TERMINOLOGY

absorption	electromagnetic radiation
accessory pigment	etiolation
aerobic	fluorescence
anaerobic	oxidation
ATP	photosynthesis
carbon dioxide	reduction
carotenes	respiration
carotenoids	visible light
chlorophyll	wavelength
chloroplast	xanthophylls
chromatography	

INTRODUCTION TO PHOTOSYNTHESIS AND RESPIRATION

In photosynthetic organisms, light energy captured from the sun is converted to chemical energy. Most of the energy that enters the living systems of Earth is through the process of **photosynthesis**, during which **carbon dioxide** and water and light energy are ultimately converted to carbohydrate and oxygen. During **respiration**, carbohydrates are oxidized to carbon dioxide and water, releasing the stored chemical energy required for growth and maintenance. In plants, photosynthesis occurs in chloroplasts, and **anaerobic** and **aerobic** stages of respiration occur in the cytoplasm and the mitochondria, respectively.

In this exercise, we will explore several aspects of photosynthesis, including some of the properties of light, the pigments required for photosynthesis, and the requirement for both light and photosynthetic pigments for the process of photosynthesis. In addition, you will perform a practical test relying on respiration to determine if seeds are viable.

PART A LIGHT ENERGY AND PHOTOSYNTHESIS

The light energy that drives photosynthesis is a tiny part of the continuous **electromagnetic** spectrum of **radiation**, and it is the part with just the right amount of energy useful in biological systems. The colors of **visible light** result from the slightly different **wavelengths** of the visible light spectrum. Generally, wavelengths of electromagnetic radiation are expressed in nanometers (nm). Visible light ranges from around 380 to 750 nm. In this part of the exercise, you will examine the composition of visible light and some of the properties of the photosynthetic pigment, chlorophyll.

MATERIALS

- a prism or spectroscope
- colored filters
- spinach leaves
- blender
- 95% ethanol
- small beaker
- watch glass
- hotplate
- square bottle
- prepared chlorophyll extract

PROCEDURE

1. Composition of Visible Light

   Using a prism or spectroscope, examine the spectrum of sunlight or light from an incandescant light bulb. Describe the colors revealed and their relationship to one another. Now, place a colored filter between the source of light and the spectroscope and describe the changes that take place in the spectrum. You should be able to relate the color of the filter to the colors absorbed or transmitted.

2. Absorption of Light by Chlorophyll

Quickly chop one or two handfuls of spinach leaves in a blender. Remove the leaves to a beaker and cover with 95% ethanol. Cover the beaker with a watch glass and gently bring the solution to a low boil by heating on the hotplate. Be careful with the boiling alcohol; it can catch fire. After the solution has turned a light green, turn off the hotplate and let the beaker and solution cool. Once it is cool, pour the solution into the square bottle. Place the bottle between the light source and the spectroscope. If the solution absorbs too much light, it can be diluted with additional alcohol. Describe the changes in the spectrum when compared with white light. Look for dark bands in the spectrum indicating regions of light **absorption** by the pigment extract.

3. Chlorophyll Fluorescence

A concentrated chlorophyll extract will be provided by your instructor (or you can prepare an extract as above, but boil the leaves for a longer period of time to increase the extraction of chlorophyll from the leaves). Pour some of the concentrated extract into a test tube and hold the test tube in front of a strong beam of incandescent light. You may have to turn the tube or change its angle, but a dark red color instead of green should become visible. When photosynthetic pigments absorb light, electrons are boosted to a higher energy level. As the electrons return to a lower energy level the energy may be emitted as light energy of a longer wavelength (the longer the wavelength, the lower the energy) or the energy may be captured in chemical bond formation as during photosynthesis. The energy may also be given off as heat. The deep red color visible in the test tube is an indication of the electrons returning to a lower energy level and emitting longer wavelenth light—the deep red color. This is known as **fluorescence**.

PART B   PHOTOSYNTHETIC PIGMENTS

The first step in photosynthesis is the absorption of light by photosynthetic pigments. Photosynthetic pigments include **chlorophylls**, which primarily absorb violet, blue, and red wavelengths and reflect green. Therefore, leaves with chlorophylls appear green. The essential form of chlorophyll is designated chlorophyll a. Chlorophyll b is also present in plants and some algae. In plants, chlorophyll b and the carotenoids are **accessory pigments** that pass their energy to chlorophyll a. There are two groups of **carotenoids**: **carotenes** and **xanthophylls**. Carotenoids appear red, orange, or yellow, but their appearance in a healthy leaf is usually masked by the more abundant chlorophylls. The chlorophylls and carotenoids are fat-soluble pigments located in the thylakoid membranes of **chloroplasts**. In an extract of photosynthetic tissue, photosynthetic pigments can be separated and identified by **chromatography**.

MATERIALS

- prepared pigment extract (in acetone)
- chromatography paper
- chromatography jars
- small brush
- chromatography solvent (9 parts petroleum ether:1 part acetone)

PROCEDURE

Separation of Photosynthetic Pigments by Paper Chromatography

Use a clean sheet of chromatography paper. About 1 cm from the bottom edge of the paper, mark a straight line with a pencil (do not use a pen). Dip the brush in the prepared pigment extract and neatly brush pigment evenly along the pencil line. Repeat the process five to ten times, or until the line is very dark green. Add a layer of solvent about 0.5 cm deep to the chromatography jar. Place the paper, pigment side down, in the chromatography jar. Be sure the solvent does not touch the line of pigment and be sure the paper is upright. The solvent will move up the paper and take the pigment extract with it, but the components in the pigment extract will move at different rates. This property enables us to separate the pigments. Let the chromatogram run for 15–30 minutes, but keep an eye on it in case it moves quickly. Remove the paper when the solvent reaches the top of the paper. Record your results before the chromatogram fades. Carotenes will appear orange-yellow, xanthophylls are pure yellow, chlorophyll b will be blue-green, and chlorophyll a will appear grass-green. Note the position of each pigment relative to the other.

PART C NECESSITY OF LIGHT AND CHLOROPHYLL FOR PHOTOSYNTHESIS

Both light and chlorophyll are necessary for photosynthesis to occur. Evidence of photosynthetic activity can be obtained by looking for places where starch accumulates and using I_2KI to test for the presence of starch. In this portion of the experiment, we will use a plant with partially masked leaves and a plant with variegated leaves to test for the presence of starch.

MATERIALS

- bean plants with partially masked leaflets
- green, red, and white variegated *Coleus* plant
- beakers
- watch glasses
- tongs
- 95% ethanol
- two hotplates
- white plate
- I_2KI stain
- dark-grown corn and pea seedlings

PROCEDURE

1. Demonstrating Regions of Photosynthesis by the Presence of Starch

   Place a beaker containing water on one hotplate and a beaker with 95% ethanol on the other hotplate. Turn on the hotplates and bring the water and the alcohol to a boil. The experiment cannot be performed until the water and alcohol are boiling.

   Several days ago a bean plant was prepared by fastening a piece of foil to each of several leaflets, blocking entry of light into the covered tissue.

   a. Cut off a masked leaflet and remove the foil.

   b. As a control, treat a leaflet that has not been masked with foil.

   c. With tongs, place the leaflets in boiling water for 30 seconds, then transfer to hot alcohol.

   d. Cover the beaker with a watch glass and boil until the leaflets lose their green color. Turn off the hotplates.

   e. Remove the leaflets (they will be dehydrated and stiff), and transfer them to a white plate containing water. Straighten and unfold the leaves if necessary.

   f. While holding the leaves against the plate, pour off the water.

   g. Slowly add I_2KI solution to cover the leaves. Let them stand for a few minutes while the color develops.

   h. Drain off excess solution and rinse gently in water.

   A blue-black color indicates the presence of starch. The region of the leaf covered with foil should appear paler than the areas exposed to light.

   Repeat this experiment with a leaf from a green, red, and white *Coleus* plant. Inspect the upper surface of the leaf and the lower surface. You may be surprised to find red areas of the leaf have stained with starch. In these areas, chlorophyll is present but is masked by red pigments not involved in photosynthesis. Only white areas of the *Coleus* leaf remain pale.

2. Etiolated Plants

   Corn and pea seeds were germinated in the dark and the seedlings kept in the dark for several days. The plants were moved into the laboratory just before the start of class. The seedlings are spindly and pale. This type of growth is called **etiolation**. The nature of the long, spindly stems will be discussed in a later exercise on plant growth and development, but the lack of green is caused by the absence of chlorophyll. Chlorophyll does not form unless a precursor molecule, prochlorophyll, is exposed to light. Because these seedlings were not exposed to light, they have remained pale and are not manufacturing food. Speculate to the fate of the seedlings if they had remained in the dark.

PART D DETECTING RESPIRATION IN SEEDS

Several factors affect seed germination. Many angiosperm seeds enter a period of dormancy, which must be overcome before germination can proceed. Some seeds require specific environmental conditions to enter germination. (These topics will be explored in a later exercise.) However, neither of these factors is relevant if the embryo within the seed is not alive. A viable seed is the first requirement of germination. Viable seeds are seeds with a living embryo capable of germinating. If a seed is viable the embryo will be conducting respiration.

During respiration, living cells obtain chemical energy required for maintenance and growth. Respiration is the **oxidation** of sugar to carbon dioxide and water, and it proceeds in two stages: an anaerobic stage and an aerobic stage. Glycolysis is the anaerobic pathway. It occurs in the cytoplasm and produces pyruvate. If oxygen is present, respiration continues with the aerobic stage, which proceeds in two phases: the Krebs cycle and the electron transport chain. The aerobic pathway occurs in mitochondria. During the Krebs cycle, electron carriers are reduced, some **ATP** is produced, and carbon dioxide is released. The electron transport chain oxidizes the reduced electron carriers from glycolysis and the Krebs cycle, producing ATP. The aerobic stage requires oxygen because oxygen is the final electron acceptor. Overall, respiration results in the formation of carbon dioxide, water, and energy. In this exercise, we will use the properties of oxidation and **reduction** during respiration to test for viability in plant seeds.

Because seed germination is important to the agricultural industry, tests have been developed to check the viability of seed batches before the seeds are planted. One of these tests uses the dye tetrazolium, or TTC. It works because living seeds respire, and respiration can be detected by the dye. Tetrazolium is colorless when oxidized and turns red or deep pink when reduced. During respiration, the electron transport system is functioning, and any added tetrazolium will interact with the seed or grain's electron transport system and accept electrons from cytochromes. Tetrazolium is reduced when it accepts electrons.

The tetrazolium test is quick and easy to use. However, use care in handling this substance, as it is a poison. Do not get any on your skin or clothing and do not ingest it. Wash immediately if you come in contact with TTC.

MATERIALS

- bean seeds (soaked overnight in the dark)
- boiled bean seeds
- soaked corn grains (soaked overnight in the dark)
- boiled corn grains
- single-edge razor blade
- china marker
- petri dishes
- filter paper
- tetrazolium (0.1% solution of 2,3,5-triphenyl-2H-tetrazolium chloride)
- forceps

PROCEDURE

Compare viability of soaked bean seeds with boiled bean seeds and soaked corn grains with boiled corn grains. With a china marker or Sharpie® pen, label four petri dishes to indicate soaked or boiled beans or corn grains. Add filter paper to the bottom of each petri dish and moisten each paper with three or four drops of tetrazolium. Select three soaked bean seeds, remove the seed coats, and separate the two halves. Place each half on the filter paper in the appropriate petri dish. Repeat with the boiled bean seeds; placing them in their own marked petri dish. Cut three soaked corn grains in half longitudinally and place the cut surfaces on the filter paper in the petri dish marked for soaked corn grains. Repeat with the boiled corn grains. Add additional drops of tetrazolium on the seeds and grains. After 15–30 minutes, use forceps to turn the seeds and grains over. If the embryos or endosperm are red or mostly red, the seeds and grains are viable. If the embryos or endosperm are pale pink or white, the seeds and grains are dead.

QUESTIONS FOR THOUGHT AND REVIEW

1. What color is visible light? Explain. _____

_____

_____

2. Why do photosynthetic leaves appear green?

_____

_____

_____

3. What wavelengths are best absorbed by chlorophyll a? by chlorophyll b? by carotenes? by xanthophylls? _____

_____

_____

_____

4. What colors of light are absorbed by ethanol extracts of spinach leaves? _____

_____

_____

5. Is there a correlation between leaf tissue color and the presence of starch? _____

_____

6. Is there a correlation between exposure to light and the presence of starch in a plant? _____

_____

7. Where might starch be located within a plant?

_____

_____

8. What is an advantage to a plant of not producing chlorophyll if the plant is growing in the dark?

_____

_____

_____

9. What are the chief products of respiration?

_____

_____

10. How does tetrazolium indicate seed viability?

_____

_____

Name: _____

Section Number: _____

PHOTOSYNTHESIS AND RESPIRATION

1. List the colors of visible light from the most energetic to the least energetic. _____
_____

2. When visible light passes through a dense tree canopy (with many green leaves), would you expect red light, green light, or blue light to pass through the canopy? Circle all that would pass through and explain your answer. _____
_____
_____

3. Which photosynthetic pigment is essential for photosynthesis in plants? _____
_____

4. What is the function of carotenoids? _____
_____
_____

5. Why is the detection of starch accumulation useful as evidence for photosynthetic activity? _____
_____

6. In the laboratory exercise, why did red regions of *Coleus* leaf tissue test positive for photosynthetic activity?
_____
_____
_____

7. Under what circumstances would etiolated growth be an advantage to a seedling? _____
_____
_____

8. What is the major difference between aerobic and anaerobic respiration? _____
_____
_____
_____

EXERCISE 8

TRANSPIRATION AND XYLEM FUNCTION

OBJECTIVES

1. Describe the role of stomata in the function of water movement within a plant.

2. Define transpiration and describe the role of transpiration in xylem water movement.

3. Describe the roles of cohesion and adhesion in the movement of water within the xylem tissue.

4. Identify the tissue and cell types involved in water movement within a plant.

5. Explain the role of the root endodermis in selective uptake of minerals from the soil.

6. Understand how active transport is involved in the uptake of minerals by roots.

TERMINOLOGY

adhesion	tension
carbon dioxide	tracheid
cohesion	transpiration
diffusion	vessel member
photosynthesis	xylem
stomata	

INTRODUCTION TO TRANSPIRATION AND XYLEM FUNCTION

One of the major obstacles plants faced when they left a watery environment and moved to a terrestrial habitat was how to obtain and conserve water. Plants must take up water from their surroundings, usually from the soil, and conserve enough water for the proper functioning of their cells and tissues. Because of **photosynthesis**, considerable water loss is inevitable. **Carbon dioxide** (CO_2) must enter leaves, and, in the process of opening leaf **stomata** to allow **diffusion** of CO_2 into leaves, plants are exposed to loss of water vapor, or **transpiration**, from their surface.

In addition to the problem of loss of water from their surfaces, plants are large and, therefore, diffusion is too slow to move water through the body of a plant. However, plants evolved to take advantage of transpiration. Plants use transpiration to pull water up their bodies, and dissolved minerals move with the water in the process. In this exercise, you will examine water loss through transpiration and see how xylem is the conduit for long-distance transport of water throughout the plant.

PART A TRANSPIRATION

For photosynthesis to occur, CO_2 must enter leaf tissue. Carbon dioxide diffuses into leaves when the stomata are open, and water vapor diffuses out of the leaves while the stomata are open. The loss of water vapor from the surface of leaves is called transpiration. Different methods can be used to measure the loss of water from plant tissues. A convenient method is with a portable autoporometer, an electronic device that measures transpiration from a leaf. In this exercise, we will use another less costly device called a potometer to observe the rate of transpiration from a branch. Also, the lifting power of transpiration will be demonstrated with a cut leafy stem and a glass capillary tube.

MATERIALS

- leafy angiosperm branches
- potometer
- thick-walled capillary tubes
- rubber tube
- ring stands and clamps
- beakers of water
- mercury (highly toxic; care must be used with this substance)
- porous clay ball

PROCEDURE

1. Measuring the Rate of Transpiration with a Potometer

   a. Your instructor will provide a simple water-filled potometer and a leafy branch (Figure 8-1). While holding the end of the branch under water, make a fresh cut in the branch and immediately attach the branch to the potometer as instructed. Look over the tubes and connections to be certain no air bubbles are present.

   b. After the branch is attached and you have checked for air bubbles, lift the open end of the potometer tube out of the water container, allowing a small air bubble to be drawn into the tube.

   c. Lower the potometer tube back into the water.

   d. The air bubble will move along the potometer tube as water is lost from the leaves through transpiration.

   e. During the course of the laboratory period, observe the movement of the air bubble. Its movement reflects the rate of transpiration of the branch.

   f. To alter the rate of movement, place a small fan near the potometer to create an air current or move the potometer set into sunlight.

2. Transpiration Lifts Water Up a Stem

   a. Take a second leafy branch and make a fresh cut under water.

   b. While the branch is under water, attach the cut end to a section of rubber tubing that is attached to a capillary tube. Take care, the capillary tube is fragile.

   c. Check to be certain that no air bubbles are present.

   d. Use your finger to cover the end of the capillary tube and place the end of the tube in a beaker of water.

   e. Use the ring stand and clamps to secure the capillary tube and branch in a vertical position.

   f. At this point, your instructor will pour a small amount of mercury into the beaker and briefly lower the capillary tube into the mercury. This

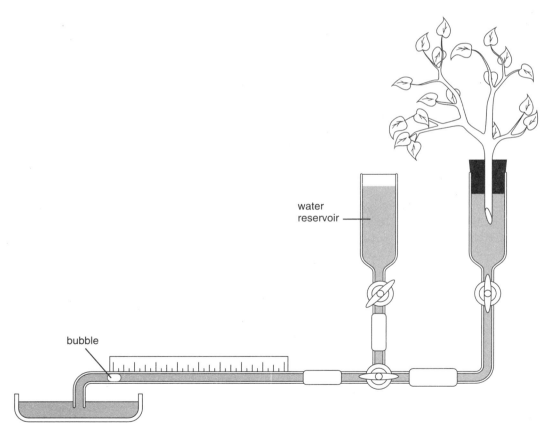

Figure 8-1 Sketch of the setup of a potometer used to study the rate of transpiration from a leafy branch.

will draw up a small bead of mercury into the tube. The progress of the mercury bead up the tube can be monitored. The progress of the bead up the tube is an indication that transpiration from the surface of the leaves is lifting the mercury bead and is dependent upon the cohesion of water molecules in the column of water.

3. Porous Ball Model of Lifting Power of Transpiration

    a. A model of the above system is displayed.

    b. Instead of a leafy branch, a hollow, water-filled clay ball is connected to a glass capillary tube full of water. The clay has microscopic, water-filled pores that allow evaporation.

    c. As before, a small bead of mercury was drawn into the tube by briefly dipping the end into the reservoir of mercury. If you don't see a mercury bead in the tube, ask your instructor to check the model.

    d. Note the movement of the mercury. Try gently fanning the clay ball to see the change in the rate of bead movement.

PART B XYLEM FUNCTION

Xylem tissue is the path of rapid water movement up the plant during transpiration. Within the xylem tissue, the water-conducting cells are **vessel members** or **tracheids**. These cell types are dead and hollow at maturity and provide natural capillary tubes within the plant. As transpiration pulls water up the vessels and tracheids, the water is under **tension** and the column is held together by the forces of **adhesion** and **cohesion**. The following portions of this exercise allow you to explore the rate of water transport through the xylem of a stem while identifying the areas of water movement within the stem and to experiment with the role of the root endodermis in regulating movement of materials into the xylem.

MATERIALS

- seedlings
- jars of eosin dye
- single-edge razor blade
- centimeter scale
- clock
- dissecting microscope
- tomato plant grown in stoppered jar containing nutrient solution
- clay or plasticine
- test tube
- sulfuric acid
- diphenylamine
- pipettes
- white ceramic test plate
- glass stirring rods

PROCEDURE

1. The Path and Rate of Water Transport in a Stem

    a. Each laboratory section will be divided into five groups of students. Each group will be responsible for measuring the rate of water transport in a plant over a given length of time. Your instructor will assign each group a time.

    b. Read through the entire procedure before beginning. Planning is essential to the success of this experiment. You will be measuring how far eosin dye climbs up a plant stem in a given length of time.

    c. Within your group decide who will perform the different parts of the experiment.

    d. Obtain a paper towel (or tape two towels together if the seedling is taller than about 20 cm). Using a scale, mark off intervals of 1 cm along the paper towel.

    e. Using a sharp single-edge razor blade, cut off the stem near its base.

    f. While holding the cut stem under water in a dish or a bucket, cut it off again 1 cm up the stem. (Why is this step desirable?)

    g. Without letting the new cut surface dry, quickly insert the cut end into a container of eosin dye solution.

    h. Immediately start timing.

    i. When the assigned length of time has elapsed, quickly remove the plant from the dye container and blot the excess dye from the end of the stem.

    j. Lay the stem on the paper towel next to the 1 cm marks and *quickly cut off the leaves*.

    k. Cut the stem into 1 cm segments starting from the top. Keep the segments in order.

    l. To find the highest segment that contains orange dye in its vascular bundles, examine each stem segment with the dissecting microscope. The orange dye may appear brownish-orange in the vascular bundles, and some of the bundles may conduct dye to a higher level than other bundles. Examine each bundle carefully, and take your measurement from the highest segment with at least one bundle containing eosin dye. Remeasure this segment to determine the height in mm that the eosin dye climbed up the stem.

m. Divide the height (in mm) by the time, to get a speed of dye uptake in mm per minute. Record the speed in Table 8-1.

Table 8-1   Rate of Water Transport in a Seedling

Group	Assigned time	Height dye reached (mm)	Speed (mm/min)
1			
2			
3			
4			
5			
		Average speed (mm/min)	

2. Nitrate Uptake in Roots

CAUTION! One of the reagents contains concentrated acid. Avoid contact with skin or clothing. In case of spills, wash immediately with plenty of water and get the instructor's help.

a. To start, press clay or plasticine around the base of the stem to be cut.

b. Cut off the shoot 1 cm above the jar stopper.

c. Lower an inverted test tube over the stump and press it into the plasticine. The tube should not touch the stump.

d. During the lab period, a drop of xylem sap will form on top of the stem stump.

e. Draw the xylem sap into a pipette and transfer the sap to a well in a ceramic test plate.

f. Then lift the stopper of the culture jar and, with a fresh pipette, transfer a drop of nutrient solution to a second well in the test plate.

g. Add 2 drops of sulfuric acid and 2 drops of diphenylamine to each well. Mix with a glass rod.

h. The color fades in a few minutes, so watch carefully. The intensity of color is directly proportional to the amount of nitrate present.

QUESTIONS FOR THOUGHT AND REVIEW

1. Define transpiration. _____

_____

_____

2. Contrast cohesion and adhesion. _____

_____

_____

_____

3. Why is water normally lost from plants during photosynthesis? _____

_____

_____

4. What is a typical rate of transpiration in a cut branch?_____

5. If a clay ball and tube are used as a model of a transpiring plant, what part of the plant corresponds to the capillary tube? _____

_____

6. If a clay ball and tube are used as a model of a transpiring plant, what part of the plant corresponds to the clay in the clay ball?_____

_____

7. At the rate determined in the path and rate of water transport in a stem exeriment, how long will it take for water to reach the top of a 100-meter-tall tree? _____

8. In the nitrate uptake experiment, which had the higher nitrate concentration, the nutrient solution or the xylem sap? _____

9. What does the result of the nitrate uptake experiment imply about the mechanism of nitrate uptake?

_____

_____

_____

10. What is the role of the Casparian strip in mineral uptake? _____

_____

_____

_____

Name: _____

Section Number: _____

TRANSPIRATION AND XYLEM FUNCTION

1. Name one reason diffusion is not the mechanism by which water moves from tree roots to leaves.

_____

_____

2. What tissue is responsible for long-distance transport of water in plants? _____

_____

3. Define transpiration. _____

_____

_____

4. Briefly describe how transpiration lifts water up a stem. _____

_____

_____

5. During the laboratory exercise, you examined a clay-ball and capillary-tube model of transpiration. What could you add to the model to make it more realistic? _____

_____

Explain your answer. _____

_____

_____

6. How does the endodermis contribute to xylem function? _____

_____

_____

7. Based on the laboratory exercise, estimate how long it would take water to reach the top of a 25-meter-tall tree. _____

_____

_____

8. Roots take in water and minerals. What tissue keeps the minerals from leaking out of the roots?

_____

MEIOSIS AND LIFE CYCLES

OBJECTIVES

1. Identify and sketch the stages of meiosis I and meiosis II and understand the events of each stage.

2. Describe the stages of a typical gametic, zygotic, and sporic life cycle.

TERMINOLOGY

alternation of genera-	haploid
tions	meiosis I
anaphase I	meiosis II
chromatid	metaphase I
chromosome	prophase I
crossing-over	sperm cell
diploid	sporic life cycle
egg cell	sporophyte
fertilization	synapsis
gamete	telophase I
gametic life cycle	zygote
gametophyte	zygotic life cycle

INTRODUCTION TO MEIOSIS AND LIFE CYCLES

Most eukaryotic organisms, including animals and plants, undergo meiosis as part of their sexual reproductive life cycle. Meiosis is a reduction division process that produces four **haploid** cells from a **diploid** cell. The diploid condition is reinstated during fusion (**fertilization**) of a haploid **egg cell** and a haploid **sperm cell** to form a diploid **zygote**. Meiosis and fusion of **gametes** are key events in the life cycle of eukaryotic organisms. There are three basic life cycles: **gametic**, **zygotic**, and **sporic**. Animals have gametic life cycles, fungi have zygotic life cycles, plants have a sporic life cycle, and all three life cycles are found in the protistan kingdom. An understanding of meiosis and an organism's life cycle simplifies understanding the life processes of that organism. In this exercise, you will examine meiosis with clay models and be introduced to the three general life cycles. Understanding these processes will enhance the study of specific organisms in later exercises.

PART A   THE STAGES OF MEIOSIS

Cells that undergo meiosis go through two nuclear divisions that produce four progeny cells from a single parental cell. Meiosis is called reduction division because each new cell has half the number of **chromosomes** as the parental cell. A reduction division that produces four cells occurs because the parental cell chromosomes replicate prior to the start of meiosis. Each chromosome has two identical and joined sister **chromatids**. During **meiosis I**, pairs of chromosomes (each with two chromatids) become visible with the aid of a light microscope. The paired chromosomes are referred to as homologous chromosomes and the process of pairing is called **synapsis**. The ploidy of a nucleus with paired chromosomes is 2n or diploid. During **prophase I**, homologous chromosomes may undergo **crossing-over**, the time when they may exchange genetic material and produce chromosomes with new genetic combinations. Before attempting to understand the process of meiosis, and because the terminology is the same, review the stages of mitosis covered in Exercise 2. For this part of the exericse, use strips of modeling clay to represent chromosomes and refer to Figure 9-1, page 64 for guidance in preparing the models. Assume your model cell has two pairs of chromosomes and take the chromosomes through meiosis I and II.

MATERIALS

■ two different colors of nondrying modeling clay

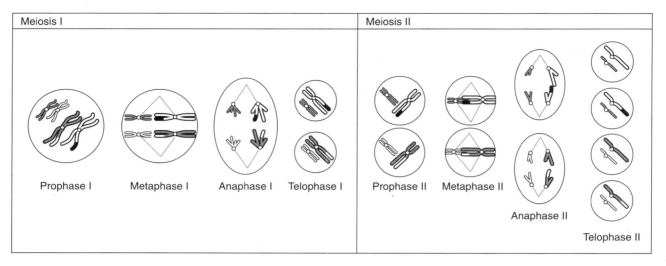

Meiosis I	Meiosis II
Prophase I Metaphase I Anaphase I Telophase I	Prophase II Metaphase II Anaphase II Telophase II

Figure 9-1 The stages of meiosis. Note that one diploid cell gives rise to four haploid cells.

PROCEDURE

1. Meiosis I

   a. **Prophase I**. Prepare strips of clay as models of chromosomes. Start with prophase I when each chromosome consists of two identical chromatids connected by a centromere. Use two strips of clay of the same color to represent sister chromatids and use two strips of clay of different colors to represent the sister chromatids of the homologous chromosome. Pair the homologous chromosomes and move small bits of clay from one chromosome to the other to represent crossing-over.

   b. **Metaphase I**. Align the paired chromosomes along either side of the theoretical cell equator.

   c. **Anaphase I**. Move the the homologous pairs of chromosomes away from each other and away from the equator toward the opposite poles of the cell, keeping the sister chromatids together.

   d. **Telophase I**. One set of the paired chromosomes is now located at each of the two poles of the cell. Nuclear division will proceed, isolating the homologous chromosomes from one another. There are now two nuclei and each nucleus is haploid. However, no chromosome replication occurs between meiosis I and II. Depending on the species, cytokinesis may or may not occur between the stages of meiosis.

2. Meiosis II

   The stages of **meiosis II** are essentially identical to mitosis. Move the chromosomes of the two nuclei through prophase II, anaphase II, metaphase II, and telophase II. You should end with four nuclei, each with chromosomes consisting of a single chromatid.

PART B GENERAL LIFE CYCLES

There are three basic types of life cycles: gametic, zygotic, and sporic. In both the gametic and the zygotic life cycles, there is one adult generation. In the gametic life cycle, the single adult generation is diploid; in the zygotic life cycle, the adult is haploid. Animals and many protists have gametic life cycles. In organisms with gametic life cycles, meiosis produces the haploid gametes, egg and sperm cells, which immediately fuse to form a diploid zygote that matures into the adult form. The gametes are the only haploid cells, and the adults are diploid.

In contrast, fungi and many algae have zygotic life cycles. In organisms with a zygotic life cycle, the gametes are produced by mitosis. After the haploid gametes fuse to produce a zygote, the zygote undergoes meiosis, producing haploid spores, which mature into the adult form. The zygote is the only diploid cell, and the adults are haploid.

Plants and some algae have a much more complex life cycle, the sporic life cycle. In organisms with a sporic life cycle, there are two adult forms, the **gametophyte** and the **sporophyte**. The gametophyte is a haploid adult; the sporophyte is a diploid adult. Both adult forms occur during a complete sporic life cycle, which also is known as an **alternation of generations**.

PROCEDURE

1. The Gametic and the Zygotic Life Cycles

Refer to Figure 9-2 to familiarize yourself with the simple gametic and zygotic life cycles. Note where meiosis occurs and what cell types are produced by meiosis.

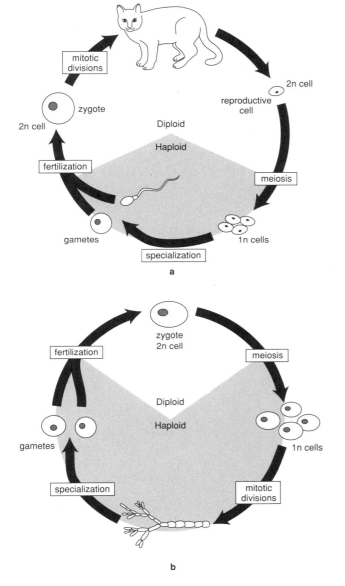

Figure 9-2 (**a**) Sketch of an idealized gametic life cycle. (**b**) Sketch of an idealized zygotic life cycle.

2. The Sporic Life Cycle

All plants have a sporic life cycle, but the size of the gametophyte relative to the sporophyte is variable. There has been an evolutionary trend toward an increasingly larger and more complex sporophyte generation. Bryophytes are the simplest plants, and in these organisms the gametophyte is the dominant or conspicuous generation. In seed plants the other extreme is seen with large, dominant sporophytes and relatively inconspicuous gametophytes.

Study the sketch of the sporic life cycle, Figure 9-3, and become familiar with the stages of this alternation of generations. Taking the time to understand the sporic life cycle now will facilitate understanding the remaining laboratory exercises during this course. Several points are essential: The gametophyte stage always begins with a spore that is produced by meiosis. This cell is sometimes referred to as a meiospore. Thus, all cells of the gametophyte form of the plant are haploid. The gametophyte generation produces gametes by mitosis (a major difference from animals). The gametes fuse to form a zygote, which is a diploid cell, and the first cell of the new sporophyte generation. Mitosis produces a diploid, multicellular sporophyte.

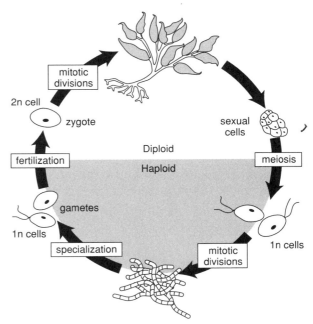

Figure 9-3 Sketch of an idealized sporic life cycle.

QUESTIONS FOR THOUGHT
AND REVIEW

1. Why is meiosis called a reduction division?

_____

_____

_____

2. Why isn't mitosis considered a reduction division? _____

_____

_____

3. What key events occur during prophase I?_____

_____

_____

4. Describe the difference between a chromosome and a chromatid. _____

_____

_____

_____

5. What is meant by the terms cell equator and cell poles? _____

_____

_____

6. Describe two major differences between the gametic and the zygotic life cycles._____

_____

_____

_____

_____

7. In reference to life cycles, what is meant by fertilization?_____

_____

_____

_____

8. Without looking at Figure 9-3, sketch the stages of a sporic life cycle. Repeat the process until you can quickly and accurately sketch all of the stages.

9. Why do you think a gametophyte is called a gametophyte and a sporophyte is called a sporophyte?_____

_____

_____

_____

10. If mammals had a sporic life cycle, what type of reproductive cells would the adult diploid organism produce? _____

_____

_____

EXERCISE 9
LABORATORY QUIZ

Name: _____

Section Number: _____

MEIOSIS AND LIFE CYCLES

1. Why is meiosis referred to as a reduction division? _____

_____

_____

2. What is the difference between a diploid and a haploid cell? _____

_____

_____

3. What general types of cellular events occur during interphase? _____

_____

4. Describe the appearance of a cell nucleus during prophase I. _____

_____

_____

5. Contrast the appearance of chromosomes during metaphase I and metaphase II. _____

_____

_____

_____

6. What events occur in a cell during telophase II? _____

_____

_____

7. List the three basic types of life cycles.

    a. _____

    b. _____

    c. _____

8. What type of life cycle occurs in animals? _____

_____

_____

In plants? _____

_____

_____

FLOWER STRUCTURE AND DIVERSITY

OBJECTIVES

1. Distinguish between terminal and axillary flowers.

2. Identify sepals, petals, stamens, and carpels within a flower.

3. Distinguish vegetative apical meristems from floral apical meristems.

4. Identify a flower as coming from a monocot or a dicot angiosperm.

5. Describe a flower as to number of parts, fusion of parts, symmetry, ovary position, completeness, and whether perfect of imperfect.

6. Describe the differences among a solitary flower and a raceme, an umbel, or a head type of inflorescence.

TERMINOLOGY

androecium	ovary
anther	ovule
calyx	pedicel
carpel	peduncle
complete flower	perfect flower
corolla	perianth
dioecious	petal
filament	pistil
fusion	raceme
gynoecium	receptacle
head	sepal
hypanthium	stamen
imperfect flower	stigma
incomplete flower	style
inferior ovary	superior ovary
inflorescence	umbel
locule	whorl
monoecious	

INTRODUCTION TO THE FLOWER

Not all plants evolved flowers, for example, ferns and pine trees. Although both of these taxa reproduce sexually, pine trees produce seeds but not flowers and ferns produce neither seeds nor flowers. (More in later exercises on how these plants accomplish sexual reproduction.) Flowers are the hallmark of the Magnoliophyta, the flowering plants. This group is informally known as the angiosperms, which means "vessel seeds." The vessel containing the seeds is located in the flower.

A flower is produced from a terminal or an axillary meristem. In the process of forming a flower, an apical meristem undergoes physiological and anatomical processes that change it from a source of indeterminte growth to a determinate structure. The stalk bearing a flower is called a **peduncle**, and the end of the peduncle bearing the floral organs is called the **receptacle**.

The typical flower is composed of four distinct **whorls** of organs: **sepals**, **petals**, **stamens**, and **carpels**. Stamens and carpels contain the male and female reproductive cells, respectively. It is the carpel that is the "vessel." Seeds are produced within the carpel. There is enormous variety in the form of flowers, and we will explore examples of typical flowers and some of this variation in structure in this exercise. Exercises follow that cover specific processes associated with the flower, including formation and delivery of pollen, the development of seeds, the formation of fruits, and the dispersal of seeds.

PART A BASIC FLOWER STRUCTURE

The four whorls of organs of flowers develop in a specific order (Figure 10-1, page 70). Sepals are first, forming the outermost whorl and serving to protect the developing flower bud. Petals form the next whorl and are usually conspicuously colored to attract pollinators. Collectively, sepals form the **calyx**, petals form the **corolla**, and sepals and petals together are the **perianth**. The third whorl consists of stamens. Each stamen is composed of a slender stalk called a **filament**

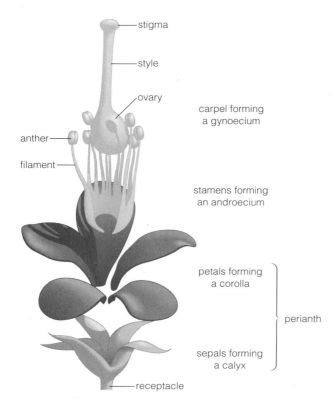

stigma

style

ovary

carpel forming
a gynoecium

anther

filament

stamens forming
an androecium

petals forming
a corolla

perianth

sepals forming
a calyx

receptacle

Figure 10-1 Flower diagram showing the whorls of organs.

that is topped by a pollen-producing **anther**. The collective term for stamens is **androecium**. The fourth and last whorl produces carpels. A carpel is a leaflike organ enclosing one or more **ovules**. Carpels are differentiated into a **stigma**, **style**, and **ovary**. The stigma captures pollen and the style directs the growth of the pollen tubes to the ovules within the ovary. (Ovules are potential seeds.) This whorl is also called a **pistil**. A simple pistil means there is a single carpel within a single flower, and a compound pistil means there are two or more carpels, separate or fused, within a single flower. The collective term for the carpels within a single flower is **gynoecium**. The cavity within an ovary is the **locule**, and the number of locules within an ovary directly corresponds to the number of carpels in the flower.

MATERIALS

- *Sedum* or other dicot flower
- prepared slide of a flower stem tip, longitudinal section
- single-edge razor blade
- dissecting microscope
- compound microscope

PROCEDURE

1. Parts of a Typical Flower

   Take a dicot flower to your workspace. First, examine the overall flower structure, locating the peduncle and receptacle at the base of the flower. Starting from the floral base, count and record the number of sepals, petals, stamens, and carpels. If necessary, remove the lower whorls to examine the upper whorls, and use a dissecting microscope to aid in seeing stamens and carpels. With a new razor blade, make a horizontal cut through the ovary region of a carpel (use a fresh flower, if needed). Using a dissecting microscope, examine the internal structure of the ovary, including the ovule or ovules, and note the number of locules. Based on the number of locules, determine if the carpel is simple or compound. Also, try to determine the number of ovules present in the flower.

2. The Apical Meristem and Flower Development

   Use a compound microscope to examine the prepared slide of a flower stem tip in longitudinal section. Note the change in the shape and extent of the apical dome during the change from a vegetative state to a developing flower, and confirm the relative locations of the sepals, petals, stamens, and carpels. The carpels are the last organs to be formed and in the process of developing use the last of the meristematic cells at the apex to form carpel tissue.

PART B VARIATION IN THE STRUCTURE OF FLOWERS

Flowers are the key source of information in determining the species of a flowering plant, and, among the different species of angiosperms, flower structure is highly variable. Recognizing differences in the structure of flowers is essential to identification of plant species. Most floral variations reflect the following differences: (1) increase or decrease in the number of parts in each whorl, (2) **fusion** of parts within or between whorls, (3) changes in the symmetry of a flower, (4) apparent position of the ovary relative to the other whorls, and (5) presence or absence of specific whorls.

Dicot flowers generally have parts in fours or fives or multiples of fours or fives; such as five sepals and ten stamens. Monocot flowers typically have parts in groups of three or multiples of three. Floral organs may exhibit different degrees of fusion; for example, sepals may be fused with one another for part or all of their length. Organs from different whorls may be partially or entirely fused with one another, such as the bases of sepals, petals, and stamens fused into a tubelike structure called a **hypanthium**, or filaments of stamens fused to the ovary of a carpel. There are two basic types of floral symmetry: regular and irregular. Regular floral symmetry is radial, irregular is not.

The position of the ovary relative to the other whorls is another consideration in flower description and identification. In a flower with a **superior ovary**, the sepals, petals, and stamens are inserted on the receptacle below the ovary. In a flower with an **inferior ovary** (and because of fusion), the sepals, petals, and stamens appear to be inserted at the top of the ovary.

Not all flowers possess all four whorls of organs. A **complete flower** has all four whorls, and an **incomplete flower** is missing one or more of the whorls. Flowers may have both sexes or just one. A **perfect flower** has both stamens and carpels and is a bisexual flower; an **imperfect flower** lacks either stamens or carpels and is a unisexual flower. Therefore, a bisexual flower lacking sepals or petals is a perfect but incomplete flower. There are two additional terms that apply to the species of plants with unisexual flowers. These plants are are either **monoecious** or **dioecious**. A monoecious species has separate staminate (male) and pistillate (female) flowers on the same individual. A dioecious species has staminate flowers on one individual and pistillate flowers on another.

An **inflorescence** is the arrangement of flowers on the floral axis. Flowers may be solitary at the terminal or axillary position or they may occur in clusters of varying architecture. Examples of simple inflorescences include the **raceme**, the **umbel**, and the **head**. In a raceme inflorescence, there is a series of solitary flowers on the floral axis. All of the flowers share the same peduncle, but each flower is attached to the peduncle by individual short stalks called **pedicels**. In an umbel in florescence, several different peduncles emerge from the same axis. The head inflorescence consists of many individual flowers, without peduncles, emerging from the same broad receptacle surface. In this portion of the exercise, examine the available solitary flowers and the floral inflorescences. Compare these flowers with the typical, basic flower just examined.

MATERIALS

- several different species of solitary flowers
- flowers exhibiting raceme, umbel, and head inflorescence types
- dissecting microscope

PROCEDURE

1. Solitary Flowers

   Examine each of the flowers designated as a solitary flower. Some of the flowers may be so small that a dissecting microscope is required to see the floral traits. Identify and count the number of sepals, petals, stamens, and carpels. To determine the number of locules per ovary, make a cross section through the ovary. Note and record the extent of fusion, the floral symmetry, and the position of the ovary. Also record whether the flower is complete or perfect or not.

   a. **Genus:** _____
      Monocot or dicot: _____

Part or feature	Number of parts
sepals	_____
petals	_____
stamens	_____
carpels	_____
locules	_____

Characteristic	
fusion of similar parts	_____
fusion of dissimilar parts	_____
symmetry	_____
ovary position	_____
complete or incomplete	_____
perfect or imperfect	_____

   b. **Genus:** _____
      Monocot or dicot: _____

Part or feature	Number of parts
sepals	_____
petals	_____
stamens	_____
carpels	_____
locules	_____

Characteristic	
fusion of similar parts	_____
fusion of dissimilar parts	_____
symmetry	_____
ovary position	_____
complete or incomplete	_____
perfect or imperfect	_____

c. **Genus:** _____
   Monocot or dicot: _____

Part or feature	Number of parts
sepals	_____
petals	_____
stamens	_____
carpels	_____
locules	_____

   Characteristic

fusion of similar parts	_____
fusion of dissimilar parts	_____
symmetry	_____
ovary position	_____
complete or incomplete	_____
perfect or imperfect	_____

d. **Genus:** _____
   Monocot or dicot: _____

Part or feature	Number of parts
sepals	_____
petals	_____
stamens	_____
carpels	_____
locules	_____

   Characteristic

fusion of similar parts	_____
fusion of dissimilar parts	_____
symmetry	_____
ovary position	_____
complete or incomplete	_____
perfect or imperfect	_____

e. **Genus:** _____
   Monocot or dicot: _____

Part or feature	Number of parts
sepals	_____
petals	_____
stamens	_____
carpels	_____
locules	_____

   Characteristic

fusion of similar parts	_____
fusion of dissimilar parts	_____
symmetry	_____
ovary position	_____
complete or incomplete	_____
perfect or imperfect	_____

f. **Genus:** _____
   Monocot or dicot: _____

Part or feature	Number of parts
sepals	_____
petals	_____
stamens	_____
carpels	_____
locules	_____

   Characteristic

fusion of similar parts	_____
fusion of dissimilar parts	_____
symmetry	_____
ovary position	_____
complete or incomplete	_____
perfect or imperfect	_____

2. Variations in Inflorescence Types

   Different species of flowers exhibiting the raceme, umbel, or head types of inflorescence are available for study with the dissecting microscope. Compare the three architectural types and record the same data asked about the solitary flowers.

   a. **Genus:** _____
      Monocot or dicot: _____
      Inflorescence type: _____

Part or feature	Number of parts
sepals	_____
petals	_____
stamens	_____
carpels	_____
locules	_____

      Characteristic

fusion of similar parts	_____
fusion of dissimilar parts	_____
symmetry	_____
ovary position	_____
complete or incomplete	_____
perfect or imperfect	_____

b. Genus: _____

Monocot or dicot: _____

Inflorescence type: _____

Part or feature	Number of parts
sepals	_____
petals	_____
stamens	_____
carpels	_____
locules	_____

Characteristic	
fusion of similar parts	_____
fusion of dissimilar parts	_____
symmetry	_____
ovary position	_____
complete or incomplete	_____
perfect or imperfect	_____

QUESTIONS FOR THOUGHT AND REVIEW

1. Indicate which of the following phrases refers to monocot flowers and which refers to dicot flowers: flower parts typically in groups of three or multiples of three? _____ flower parts typically in groups of four or five? _____

2. What is a complete flower? _____

_____

_____

3. What is a perfect flower? _____

_____

_____

4. Describe the difference between a superior and an inferior ovary. _____

_____

_____

_____

5. Briefly describe two different types of flower inflorescences. _____

_____

_____

_____

6. What is the function of sepals? _____

_____

_____

7. What is the function of petals? _____

_____

_____

8. Describe the difference between a flower with regular symmetry and one with irregular symmetry.

_____

_____

_____

_____

9. Can stamens be fused to ovaries? _____

Explain. _____

_____

_____

10. How many different types of flowers can be found on a monoecious plant? _____

EXERCISE 10
LABORATORY QUIZ

Name: _____

Section Number: _____

FLOWER STRUCTURE AND DIVERSITY

1. What is the primary function of a flower? _____
    _____
    _____

2. List in order from first formed, the four basic organs within a typical flower.

    a. _____

    b. _____

    c. _____

    d. _____

3. List the parts of an ovary.

    a. _____

    b. _____

    c. _____

4. What is the function of an ovule? _____
    _____
    _____

5. Contrast a perfect flower with an imperfect flower. _____
    _____
    _____

6. What happens to the shoot apical meristem during flower development? _____
    _____
    _____

7. Describe an advantage to a plant in having inferior ovaries. _____
    _____
    _____

8. What is the function of the style? _____
    _____
    _____

EXERCISE 11

POLLEN DEVELOPMENT AND POLLINATION

OBJECTIVES

1. Understand the stages of microsporogenesis and pollen grain development.

2. Understand the stages of microgametophyte development and maturation.

3. Describe the stages of pollination and pollen grain germination.

4. Identify specific flower traits associated with specific pollinators.

TERMINOLOGY

allogamous	monosulcate
androecium	nectary
anemophilous	ornithophilous
anther	palynology
autogamy	pollen grain
cantharophilous	pollen tube
chiropterophilous	pollination
cleistogamous	pollination syndromes
entomophilous	pollination vector
extine	sperm cell
filament	sporopollenin
generative cell	stamen
intine	stigma
microgametogenesis	stomium
microgametophyte	tapetum
microsporangium	tricolpate
microspore	triporate
microsporocyte	tube cell
microsporogenesis	

INTRODUCTION TO POLLEN

Collectively, the male whorl of organs within a flower is the androecium. The androecium is composed of stamens, and stamens are composed of two parts, the filament and the anther. An anther typically encloses four **microsporangia**, which are the sites of pollen production. The first stage in pollen development is **microsporogenesis**, which results in the production of **microspores** from **microsporocytes** by the reduction-division process, meiosis. Each microspore then undergoes mitosis and produces a two-celled **pollen grain**. The resultant two cells within the microspore wall are the **generative cell** and the **tube cell**. The generative cell will undergo an additional mitotic division to produce two **sperm cells**. A pollen grain is an immature male gametophyte; therefore, male gametophyte development begins inside a microspore wall and while still retained within a microsporangium. Upon release from the anther, the pollen grain is either a two-celled (binucleate pollen) or, less commonly, a three-celled (trinucleate pollen) immature male gametophyte (or **microgametophyte**).

Angiosperm pollen has two cell-wall layers, the **intine** and the **extine**. The intine, next to the living cell, is mostly cellulose, but the outer extine is composed of complex polymers, including **sporopollenin**. Sporopollenin is a polymer of carotenoids and extremely tough, protecting the pollen grain from desiccation and microbial attack. The extine of most pollen grains is highly sculptured in a species-specific pattern. There are thin areas in the extine, pores or furrows, through which pollen tubes emerge during pollen grain germination. The sculptured wall and the size of the grain are useful traits in identifying plant species. Generally, pollen grains of monocots and some dicots have a single, long furrow along one side of the grain. This is called **monosulcate** pollen. **Tricolpate** or **triporate** pollen, with three furrows or three pores, is typical of most dicots. Because pollen fossilizes so well, fossil pollen grains are particularly useful in determining the species of plants that lived during past geological periods. In addition to sporopollenin, some pollen grains contain compounds that cause allergic reactions (hay fever or pollen allergies) in sensitive individuals.

After pollen grains are released from anthers and are deposited on the **stigma**, they continue to mature. **Pollination** is the process of pollen grain transfer to a receptive stigmatic surface of a carpel. Fertilization occurs later, after maturation of the male gametophyte. If a pollen grain is deposited on a compatible carpel, the pollen grain will germinate and a **pollen tube** will grow down the carpel style. During this process, if it has not already done so, the generative cell divides to produce the two sperm cells. The pollen tube efficiently delivers the sperm cells directly to the female gametophyte.

The study of plant pollen, **palynology**, is useful in such diverse areas as paleobotany, petroleum exploration, and authenticating honey sources (pollen is an important food source for honey bees). The development of pollen and its delivery to the carpel are featured in this exercise.

<table>
<tr><td>**PART A**</td><td>**DEVELOPMENT OF THE POLLEN GRAIN**</td></tr>
</table>

The monocot genus *Lilium* is typically used in introductory plant biology classes to study the development of pollen grains within anthers (Figure 11-1). *Lilium* flowers have six stamens, and the anther of each stamen has four microsporangia. The microsporangia are also called pollen sacs because they are the sites of pollen development. Within each pollen sac, a nutritive layer of cells, the **tapetum**, surrounds the cells that will become pollen grains. The cells of anthers are diploid, but the cells within the pollen sacs that are destined to become spores will undergo meiosis; these are the microsporocytes. The process that produces haploid microspores from diploid microsporocytes is

called microsporogensis (the origin of microspores). After meiosis is complete, each of the four haploid microspores produced by a microsporocyte divides by mitosis and differentiates into a two-celled pollen grain. One cell within the pollen grain is the generative cell and the second is the tube cell. A mature pollen grain is an immature male gametophyte. The remaining steps in **microgametogenesis** occur after pollination. This portion of the exercise illustrates the changes that occur within the *Lilium* pollen sac as microsporocytes undergo meiosis and differentiate into pollen grains.

MATERIALS

- fresh lily flowers
- prepared slide of *Lilium* anther, cross section, early prophase
- prepared slide of *Lilium* anther, cross section, first division
- prepared slide of *Lilium* anther, cross section, pollen tetrads
- prepared slide of *Lilium* mature anthers, cross section
- single-edge razor blade
- compound microscope

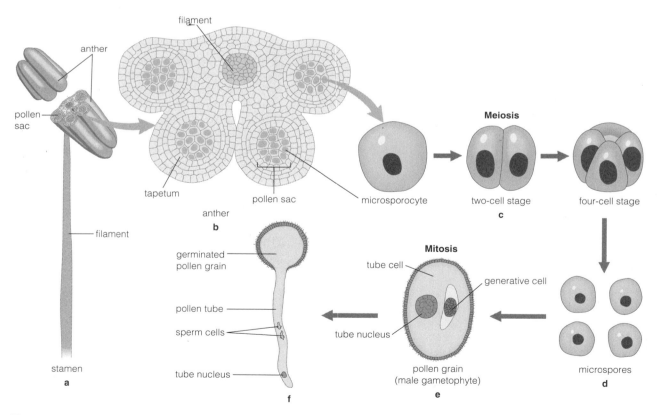

Figure 11-1 Development of pollen grain. (**a**) Stamen. (**b**) Cross section of anther. (**c**) Meiosis of microsporocyte produces four microspores. (**d**) Microspores. (**e**) Mature pollen grain. (**f**) Mature microgametophyte.

PROCEDURE

1. **Lily Anthers**

   Remove an entire stamen from a living lily flower. Examine the external morphology of the bilobed anther. Note the site of attachment of an anther to its filament; hold the base of the filament and gently shake the stamen to see how the anther moves on its filament. Under a dissecting microscope, use a sharp razor blade to cut the anther in half crosswise. Look at the cut surface and count the number of visible chambers. Set aside the cut anther for further study.

2. *Lilium* **Anthers in Early Prophase I of Microsporogenesis**

   First, on low power of a compound microscope, locate an anther cross section on a prepared slide of *Lilium* anthers in early prophase (see Figure 11-1b). The bilobed nature of lily anthers and a cross section through the filament should be clearly visible. (Note the vascular tissue within the anther and the filament. You should be able to identify xylem and phloem tissues.) Count the number of chambers in each anther cross section; these are the pollen sacs, the sites of pollen development. Within each pollen sac is a cluster of cells undergoing early stages of meiosis. These are the microsporocytes. Change to a higher power objective lens and closely examine several microsporocytes within the pollen sac. Most are in early prophase I; the nucleoli are still visible, but the chromosomes are condensing.

3. *Lilium* **Anthers in Meiosis II of Microsporogenesis**

   Obtain a second slide of a *Lilium* anther labeled as second division. Locate a pollen sac with the microscope and compare the cellular contents with the previous slide. Keep in mind that some of the developing microspores will be seen in polar view and others in equatorial view. Locate an equatorial view and note that two nuclei with an early cell wall separating them are present within each microsporocyte cell. The chromosomes are still condensed and nucleoli are not visible.

4. *Lilium* **Anthers in the Pollen Tetrad Stage of Microsporogenesis**

   After locating the pollen sacs, use the high power objective lens to study the tetrads of microspores. Locate cells with three or four nuclei and developing cell walls between the nuclei. Chromosomes are still condensed. When cytokinesis is complete, the microspores will separate and continue development, becoming pollen grains.

5. *Lilium* **Mature Anthers with Pollen Grains**

   Locate pollen sacs and examine the mature pollen grains at intermediate and high power. Note that pollen grains are separate and the wall pattern is visible. Lily pollen is binucleate when released. Each pollen grain contains two cells; the generative cell and the tube cell. The nuclei of these cells are visible within each pollen grain. Note that the tapetum is now gone and two weak regions in each anther are developing between adjacent pollen sacs. Each of these regions (a **stomium**) opens to shed pollen grains.

PART B POLLEN GRAIN GERMINATION

In *Lilium* the pollen grain is shed from the anther at the two-celled, or binucleate, stage. When released from the anther, pollen grains are actually immature microgametophytes; maturation occurs after pollination. When a pollen grain germinates on a compatible flower stigma, a pollen tube emerges. The pollen tube nucleus directs the growth of the pollen tube, and the generative cell divides to produce sperm. Once sperm are present within the pollen tube, the microgametophyte is mature. In this portion of the pollen exercise, you will examine the growth of living pollen tubes and examine pollen tubes within a flower style.

MATERIALS

- *Impatiens* plants with mature pollen-bearing flowers
- sucrose solution in a dropper bottle
- prepared slide of *Lilium* stigma and pollen tubes
- compound microscope
- depression slides
- coverslips

PROCEDURE

1. **Germination of Living Pollen Grains**

   At the beginning of the lab session, place fresh pollen in a drop of sucrose solution on a depression slide. Add a cover slip and view with a compound microscope. Periodically during the laboratory, look at the pollen grains. Watch for pollen grain germination and pollen tube elongation.

2. Pollen Tube Growth

Examine the *Lilium* prepared slide with pollen tubes and stigma. Dozens of pollen grains are on the surface of the stigma and many grains have germinated. The hollow center of the style if filled with elongated pollen tubes. Try to locate nuclei along the length of pollen tubes.

PART C POLLINATION

The pollen grain is a remarkable structure; its evolution led to the advent of direct delivery of male gametes to female gametes. Pollination is the transfer of pollen from stamen to stigma, **pollination syndromes** are the features that promote a successful transfer of pollen, and **pollination vectors** are what actually make the transfer. Pollination vectors may be abiotic or biotic. Some flowers self-pollinate; this is **autogamy**. If a flower bud remains closed prior to self-pollination, the flower is **cleistogamous**. Flowers requiring cross pollination are **allogamous**. In this part of the laboratory exercise, you will examine displays of pollination syndromes and the vectors involved in the represented syndromes.

MATERIALS

- wind-pollinated flowers
- representative flowers pollinated by beetles, flies, moths, butterflies, bees, hummingbirds, and bats

PROCEDURE

1. Wind: An Abiotic Vector

   Wind is a major vector of pollination in the northern hemisphere. Technically, wind-pollinated flowers are **anemophilous** flowers. Examples of plants that are wind-pollinated include grasses and oak trees. Anemophilous flowers tend to look very different from animal-pollinated flowers. For example, there is no need for colorful petals and these flowers frequently lack sepals and petals altogether. Examine the flowers on display and note any features common among the different species. In particular, note which floral whorls are present.

2. Biotic Vectors

   Many features of flowers serve to recruit animals as pollination vectors and seed dispersers. A general term for insect-pollinated flowers is **entomophilous**. Beetle-pollinated flowers have the

special term **cantharophilous**. **Ornithophilous** flowers are pollinated by hummingbirds, and bats pollinate **chiropterophilous** flowers. Each flowering plant species benefits from having unique identifying features and features that reward its pollinator. As a result, flowering plants have evolved an enormous variety of flower colors, shapes, scents, and edible parts. An example of an edible reward for pollinators is the floral **nectary**. Nectaries are sources of a sugary liquid attractive to many pollinators. Some plant and pollinator relationships are so specialized that the plant can only be pollinated by a specific species of animal.

Inspect the available flowers demonstrating pollinator preferences. As you examine the flowers, keep in mind the following animal traits:

a. Beetles are clumsy fliers.

b. Bees are nimble fliers and attracted to yellow, blue, and bee-purple (an ultraviolet wavelength not visible to humans).

c. Flies are attracted to the odor and color of rotting flesh or dung.

d. Hummingbirds have excellent vision, can hover, and have long tongues for gathering nectar.

e. Moths and butterflies also have long tongues for gathering nectar but cannot hover.

f. Bats generally fly at night and have an excellent sense of smell.

Record specific features that appear to be associated with pollination by the listed animals.

Pollinating Animal	Flower Trait
beetle	_____
	_____
	_____
fly	_____
	_____
	_____
moth	_____
	_____
	_____
butterfly	_____
	_____
	_____

Pollinating Animal	Flower Trait
bee	_____
	_____
	_____
hummingbird	_____
	_____
	_____
bat	_____
	_____
	_____

QUESTIONS FOR THOUGHT AND REVIEW

1. How many pollen sacs are in each anther?

_____

2. What features can be used to identify the stages of microsporogenesis within a pollen sac? _____

_____

_____

3. At the end of meiosis, why are microspores attached to one another in tetrads?_____

_____

_____

4. How many nuclei does each microspore contain? _____ What is their ploidy? _____

_____

_____

5. How many nuclei does each pollen grain contain? _____ What is their ploidy? _____

_____

_____

6. At the stage of pollen grain release from anthers, the pollen grains are immature microgametophytes. Why? _____

_____

_____

7. What cells could be found within a mature pollen tube?_____

_____

_____

8. Fill in Table 11-1 with an appropriate flower trait.

Table 11-1 Flower Traits Associated with Specific Pollination Vectors

Trait	Beetle	Bee	Butterfly	Bird	Wind
Petal color					
Flower shape					
Odor					
Nectar					

9. Generally, would you expect wind-pollinated flowers to be complete or incomplete?

_____

10. Of what benefit is a floral nectary to a plant species? _____

_____

_____

EXERCISE 11
LABORATORY QUIZ

Name:_____

Section Number: _____

POLLEN DEVELOPMENT AND POLLINATION

1. What part of the sporic life cycle does a pollen grain represent? _____
    _____
    _____

2. What is the function of the generative cell within a pollen grain? _____
    _____
    _____

3. What is the difference between pollination and fertilization? _____
    _____
    _____
    _____

4. Describe the function of sporopollenin in pollen grain walls. _____
    _____
    _____

5. How do pollen grains travel to ovaries? _____
    _____
    _____

6. Name the process by which diploid microsporocytes produce haploid microspores._____
    _____

7. Where would you find a mature microgametophyte in a flowering plant? _____
    _____

8. What is the purpose of the pollen tube? _____
    _____
    _____

EXERCISE 12

EMBRYO DEVELOPMENT AND SEED FORMATION

OBJECTIVES

1. From a section through the ovary of a flower, identify the visible parts of the ovary and state the value of each part in plant reproduction.

2. Understand the stages in development of the mature megagametophyte from a megasporocyte.

3. Understand the stages in double fertilization and subsequent development of the embryo from the zygote and of the endosperm from the primary endosperm nucleus.

4. Describe a seed. Know the parts of a seed and the functions and origin of the parts.

5. Describe the relationship between a seed and a fruit.

TERMINOLOGY

antipodal cell	megasporocyte
axile placentation	megasporogenesis
bisporic	micropyle
carpel	monosporic
central cell	nucellus
double fertilization	ovary
egg cell	ovule
embryo	placentation
embryo sac	polar nuclei
embryogenesis	primary endosperm
endosperm	nucleus
fruit	seed
gynoecium	seed coat
integument	suspensor
locule	synergid
megagametogenesis	tetrasporic
megagametophyte	triploid
megasporangium	zygote
megaspore	

INTRODUCTION TO THE FLOWERING PLANT MEGAGAMETOPHYTE, EMBRYO, AND SEED

Collectively, the female reproductive structures within a flower form the **gynoecium**. The gynoecium consists of one or more free or united **carpels**, each of which is composed of a stigma, style, and **ovary**. The young ovary of a carpel contains one or more **ovules**. Each ovule is a **megasporangium** surrounded by **integuments**. In flowering plants, the megasporangium is sometimes referred to as a **nucellus**, a term frequently encountered in scientific literature. Within the megasporangium, meiosis occurs and, from the products, develops a **megagametophyte**. Within the megagametophyte an **embryo** sporophyte is established surrounded by nutritive tissue and a **seed coat**. The **seed** or seeds are contained within the ovary.

PART A MEGASPOROGENESIS

The typical flowering plant ovule is a megasporangium surrounded by two integuments. A pore in the integuments, called the **micropyle**, is at one end of the ovule. During **megasporogenesis** a **megasporocyte** undergoes meiosis to produce four **megaspores**. Each ovule has one megasporangium, and there is a single, large megasporocyte within each megasporangium. The megasporocyte undergoes meiosis to produce four megaspore nuclei within the common cytoplasm of the original megasporocyte cell. One or more of the **megaspores** will continue developing to produce a megagametophyte. In this section of the exercise, you will examine ovules in a carpel and the formation of megaspores.

MATERIALS

- fresh lily flowers
- single-edge razor blade
- prepared slide of *Lilium* ovule at the megasporocyte stage
- prepared slide of *Lilium* ovule at the megaspore stage
- dissecting microscope
- compound microscope

PROCEDURE

1. The Lily Carpel

   Examine the carpel of the fresh lily flower. Locate the stigma, style, and ovary. With a single-edge razor blade, make a horizontal cut across the ovary and count the number of **locules**. Relate the number of locules to the number of fused carpels within the ovary. Note the position of the ovules within the ovary. The placement of ovules within the ovary is called **placentation**. The multicarpellate ovary in lily has separate locules for each carpel with the ovules attached at the center. This is called **axile placentation**.

2. *Lilium* Ovule with Megasporocyte

   Take a *Lilium* ovule slide with the megasporocyte stage and use low power of a compound microscope to locate an ovary cross section. Scan the ovary for an intact and complete ovule (Figure 12-1). Focus on the selected ovule and change to a higher power objective lens. A median section of the lily ovule will reveal the double integument, a simple megasporangium, and an enormous megasporocyte with a prominent nucleus. All visible cells are diploid at this stage. It is the megasporocyte that will undergo meiosis to produce megaspores during the process of megasporogenesis.

3. *Lilium* Ovule with Megaspore Nuclei

   Use low power to locate an ovule with four megaspore nuclei. It may be difficult to find a preparation showing all four megaspore nuclei in one view, but three visible megaspore nuclei will do. Switch to high power for a closer look at the ovule. The megasporocyte has undergone meiosis, producing four megaspore nuclei. In the majority of angiosperms, the megaspores undergo mitosis to produce eight nuclei.

PART B MEGAGAMETOGENESIS

One or more of the megaspores (depending on the species) enters **megagametogenesis** and participates in the production of a single megagametophyte, more commonly known as the **embryo sac** in angiosperms. If one megaspore participates in embryo sac formation, the pattern of development is said to be **monosporic**. **Bisporic** development involves two megaspores and **tetrasporic** involves all four megaspores. Other patterns of embryo sac development are also possible within the flowering plants. Lily is commonly used to illustrate megasporogenesis and megagametogenesis in angiosperms because it produces large megagametophytes that are easy to study. (Although the type of embryo sac development in lily is actually tetrasporic, lily is usually illustrated in beginning botany textbooks as though it is the more common monosporic type.)

In monosporic embryo sac development, the three megaspores closest to the micropylar end of the ovule degenerate. The one surviving megaspore undergoes three successive mitotic divisions to produce eight nuclei. The eight nuclei migrate within the common cytoplasm of the embryo sac; three nuclei move toward the micropylar end, three move to the opposite end, and two move to the equatorial region of the cell. Cell walls develop, establishing seven cells. The three cells at the end opposite the micropyle are called the **antipodals**. Two nuclei, called **polar nuclei**, are retained within the large **central cell**. Next to the micropyle are three cells: an **egg cell** flanked by two **synergid cells**. The three antipodal cells, two synergid cells, one egg cell, and one central cell form the embryo sac. Angiosperms do not make archegonia. In other words, the mature lily embryo sac is an eight-nucleate, seven-celled megagametophyte.

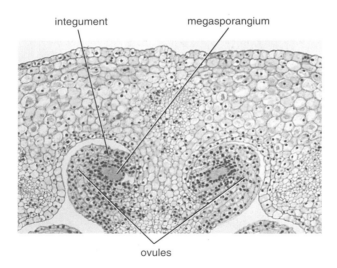

integument   megasporangium

ovules

Figure 12-1 *Lilium* ovary cross section. Two ovules visible in longitudinal section. Note integuments and megasporangia.

MATERIALS

- prepared slide of *Lilium* ovule with a mature embryo sac
- compound microscope

PROCEDURE

Lilium Ovule with Mature Embryo Sac

Scan the ovary cross section at low power and look for an embryo sac with eight nuclei (Figure 12-2). A complete embryo sac will be difficult to find, and you may have to settle for viewing only seven of the eight nuclei. After mitosis, the eight nuclei migrate through the cytoplasm to different regions of the cell. Three nuclei migrate to the micropylar end of the embryo sac, three migrate to the end opposite the micropyle, and two migrate to the center. Cell walls form around each of the six nuclei at the cell ends. This produces seven cells; six have one nucleus each and the seventh

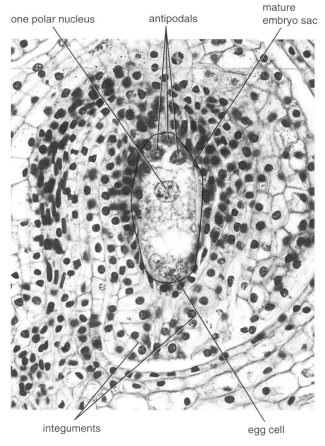

Figure 12-2 *Lilium* embryo sac, longitudinal section. A mature embryo sac contains eight nuclei and seven cells. The seven cells are the egg cell, two synergid cells, three antipodal cells, and the central cell with its two polar nuclei.

cell has two nuclei, the polar nuclei. The cell in the center of the cluster of three cells at the micropylar end is the egg cell. The two cells flanking the egg cell are synergids. The three cells at the end opposite the micropyle are called antipodal cells. The cell in the center with the two polar nuclei is the central cell. At this stage the embryo sac is a mature megagametophyte ready for fertilization.

PART C EMBRYOGENESIS AND SEED DEVELOPMENT

The pollen tube grows down the style of the carpel toward the ovules. Generally, after passing through the micropyle, the pollen tube enters one of the synergid cells, which is degenerating. One of the two sperm cells delivered by the pollen tube enters the egg cell and the other fuses with the two polar nuclei in the central cell. This process is **double fertilization**. Although double fertilization occurs in a couple of gymnosperm genera, angiosperms are the only division of plants where the products of both fertilization events continue development. The egg and sperm fertilization event creates the diploid **zygote**. The zygote will undergo numerous mitotic divisions to produce the multicellular embryo sporophyte. The polar nuclei and sperm fertilization event produces a **triploid primary endosperm nucleus**. The primary endosperm nucleus continues mitotic divisions to form a new tissue called **endosperm**. If cytokinesis follows mitosis in endosperm formation, the endosperm is cellular. If free nuclear division occurs without cell wall formation, a "liquid" endosperm is produced. Endosperm tissue is unique to angiosperms.

Embryogenesis and seed development begin after fertilization. The zygote undergoes mitotic divisions followed by cytokinesis to form the embryo of the seed. The primary endosperm nucleus undergoes mitotic divisions, sometimes followed by cytokinesis, to form the endosperm. Endosperm is the nutritive tissue of angiosperm seeds, and in most angiosperms the endosperm nuclei are triploid. As the embryo and endosperm develop, the integuments differentiate into the seed coat. A mature seed consists of an embryo, food reserves in the endosperm or the cotyledons, and a seed coat.

The anatomy of angiosperm embryos is highly variable. In a typical dicot, like *Capsella*, an embryo with two large cotyledons forms during seed development. In *Capsella* the embryo axis bends over on itself within the confines of the seed coat. Little endosperm remains when the seed is mature, and the seed coat is thin and hardened. We will use *Capsella* to study embryo and seed development in this portion of the exercise.

MATERIALS

- prepared slide of *Capsella* early embryo, longitudinal section
- prepared slide of *Capsella* early cotyledon stage of embryo, longitudinal section
- prepared slide of *Capsella* mature embryo
- compound microscope

PROCEDURE

1. *Capsella* Early Embryo

   Starting with the low power objective lens of a compound microscope, locate an ovule with an early embryo on the appropriate prepared slide of *Capsella*. Change to high power to view the details of the early embryo consisting of two parts: a ball of cells atop a stalk of cells. The ball of cells is the embryo proper, and the stalk of cells is called a **suspensor**. The suspensor pushes the embryo proper into the region of the developing endosperm. The embryo proper will continue mitotic divisions to form the mature embryo. Although the endosperm cells are not conspicuous in this preparation, look for endosperm nuclei. Mitotic divisions of the triploid primary endosperm nucleus produces the endosperm nuclei.

2. *Capsella* Embryo with Early Cotyledons

   Using the *Capsella* prepared slide indicating the presence of early cotyledons, scan at low power to find a complete embryo then switch to high power. The ball of cells of the embryo proper has grown and differentiated; the two cotyledons and the root-hypocotyl axis are now recognizable. Look for the cellular endosperm and changes in the anatomy of the cells of the integuments.

3. *Capsella* Mature Embryo

   Now study a prepared slide with a mature *Capsella* embryo (Figure 12-3). After locating an intact embryo, study the embryo at high power. Note the appearance of the cotyledons relative to the previous slide. The cotyledons are bending back on the radicle. Look for any procambium within the root-shoot axis of the embryo. Look for remains of endosperm cells and note the differentiated seed coat. The entire structure, seed coat, endosperm remains, and embryo, is now a seed. Change back to low power and note the location of this seed within a larger structure. Several

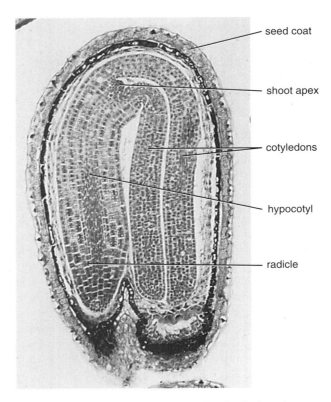

Figure 12-3 *Capsella* mature embryo, longitudinal section.

other seeds are near the one you have been examining. All of these seeds developed from individual ovules within an ovary. The ovary wall has matured into a **fruit** wall. Therefore, the seeds are enclosed within a fruit. In *Capsella*, the fruit wall is thin.

QUESTIONS FOR THOUGHT AND REVIEW

1. What is an ovule? _____

2. What is another name for the angiosperm embryo sac?_____

3. Sketch the steps in the formation of a megagametophyte.

4. How many cells are present in a mature megaga-metophyte? _____

5. In double fertilization, what is fertilized?

_____

_____

6. What is endosperm? _____

_____

What is its function? _____

_____

_____

7. Is *Capsella* a dicot or a monocot? Explain your reasoning. _____

_____

_____

8. Are the cotyledons part of the embryo? Explain.

_____

_____

_____

_____

9. What is the hypocotyl-root axis in an embryo?

_____

_____

_____

10. Explain why there are numerous mature seeds within a single fruit wall of *Capsella*. _____

_____

_____

_____

EXERCISE 12
LABORATORY QUIZ

Name:_____

Section Number: _____

EMBRYO DEVELOPMENT AND SEED FORMATION

1. Describe an ovule. _____
_____

2. What is the function of a megasporocyte?_____
_____
_____

3. What is the ploidy of a megaspore? _____
_____
_____

4. How many cells form a mature embryo sac, and how many nuclei are contained within each of these cells?
_____

5. What is the function of the synergid cells? _____
_____
_____

6. Describe the events leading to double fertilization._____
_____
_____
_____

7. What is the function of endosperm?_____
_____
_____

8. In the laboratory exercise, you examined a *Capsella* embryo. What organs of the embryo were visible within the seed? _____
_____
_____

FRUIT MORPHOLOGY AND SEED DISPERSAL

OBJECTIVES

1. Identify a fruit as simple, multiple, or aggregate, and state whether it contains accessory tissue.

2. Determine whether a fruit developed from a superior or an inferior ovary.

3. Identify the type of seed dispersal employed by specific fruits or seeds.

TERMINOLOGY

abiotic dispersal	inferior ovary
accessory fruit	legume
achene	locule
aggregate fruit	mesocarp
berry	multiple fruit
biotic dispersal	nut
capsule	ovary
carpel	pepo
caryopsis	pericarp
dehiscent	petal
drupe	pome
dry fruit	receptacle
endocarp	samara
exocarp	seed
fleshy fruit	sepal
follicle	silique
fruit	simple fruit
hesperidium	superior ovary
indehiscent	

INTRODUCTION TO FRUIT MORPHOLOGY AND SEED DISPERSAL

A **fruit** is a mature, **seed**-containing **ovary**. Two areas of focus should be kept in mind during this exercise: which floral tissues gave rise to the different parts of an individual fruit, and how the fruit promotes seed dispersal.

The number of seeds within a fruit varies from one to many thousands and depends on the plant species. As the ovules within an ovary mature into seeds, the

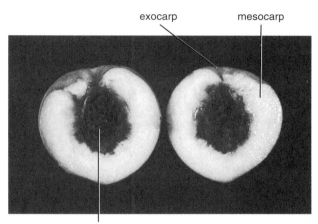

endocarp fused to seed coat

Figure 13-1 Mature peach fruit with visible layers of pericarp.

ovary wall differentiates to form the fruit wall, called a **pericarp**. The pericarp differentiates into three tissues: the **exocarp**, the **mesocarp**, and the **endocarp** (Figure 13-1).

The features of the pericarp tissue give individual fruits specific characteristics. For example, a fruit may be **fleshy** or **dry**. If dry, a mature fruit may split open (the fruit is **dehiscent**) to expose seeds. Some dry fruits are **indehiscent**. Some fruits include tissue that is not part of the ovary, such as fused **sepals** and **petals** or part of the receptacle. These fruits are referred to as **accessory fruits** because of the presence of additional, or accessory, tissue. Typically, fruits that develop from **inferior ovaries** have accessory tissue.

There are three basic categories of fruit: **simple**, **aggregate**, and **multiple**. Simple fruits develop from a single **carpel** or from several fused carpels within a single flower, aggregate fruits develop from separate carpels within one flower, and multiple fruits develop from the coalesced ovaries of more than one flower. To be consistent throughout this exercise, for fruits with accessory tissue, we will include the accessory tissue in descriptions of the pericarp.

In addition to protecting seeds, fruits serve to aid in seed dispersal. For example, fruits or seeds may be modified to promote seed dispersal by wind, water, insects, birds, or mammals. For thousands of years, humans have gathered seeds and fruits and have manipulated the remarkable diversity in seed and fruit development for food and other goods.

PART A FRUIT MORPHOLOGY

Fruits are categorized as dry (such as corn or walnuts) or fleshy (such as tomatoes or peaches); and all fruits, whether dry or fleshy, are classified as simple, aggregate, or multiple, and as either having or lacking accessory tissue. In addition, these natural characteristics have been combined to classify fruits into different types. The last classification of fruits into types is an artificial classification used by people as a convenience in discussing fruits. For example, apples are simple fruits with accessory tissue and are classified as **pomes**. Examine each fruit and look for clues to help you classify the fruit as simple, aggregate, or multiple, and determine whether accessory tissue is present or not. Also, try to determine whether each fruit differentiated from an inferior or a superior ovary. After making these determinations, use the key to classify fruit types for the indicated fruits and complete the chart below.

MATERIALS

- prepared slide of tomato fruit young ovary in cross section
- prepared slide of apple fruit in cross section
- orange fruits
- apple fruits
- strawberry fruits
- pineapple fruits
- variety of additional fruit types
- compound microscope

PROCEDURE

1. Fruit Development and Structure

   a. Tomato fruit anatomy

   Examine the prepared slide of a young ovary of tomato with low power of a compound microscope. The tomato fruit is a simple fruit without accessory tissue. Identify the number of locules, locate developing seeds, and examine the maturing ovary wall. Try to discern the layers of the pericarp.

   b. Apple fruit anatomy

   Using low power of a compound microscope, examine the prepared slide of an apple fruit.

The apple fruit is a simple fruit with accessory tissue. Locate the locules, developing seeds, ovary wall, and accessory tissue. Note the fusion of the accessory tissue with the ovary wall.

2. Key to Common Fruit Types

   Examine the fruits on display and determine whether each fruit is dry (such as corn or walnuts) or fleshy (such as tomatoes or peaches). As you study each fruit, try to visualize the flower that gave rise to the particular fruit.

   1a. *Simple fruit*. Fruit developed from a single ovary (can be one carpel or several fused carpels) of a single flower.

   　2a. Fruit fleshy at maturity.

   　　3a. Ovary superior. Pericarp formed from ovary wall.

   　　　4a. Pericarp fleshy throughout.　**Berry**

   　　　4b. Outermost layers of pericarp leathery.　**Hesperidium**

   　　　4c. Exocarp thin, mesocarp fleshy, endocarp woody.　**Drupe**

   　　3b. Ovary inferior. Pericarp formed from fused ovary and floral tube.

   　　　4a. Pericarp fleshy.　**Pepo**

   　　　4b. Exocarp and mesocarp fleshy, endocarp papery.　**Pome**

   　2b. Fruit dry at maturity.

   　　3a. Dehiscent. Fruit splitting open at maturity.

   　　　4a. Fruit developed from one carpel.

   　　　　5a. Fruit splits along one seam.　**Follicle**

   　　　　5b. Fruit splits along two seams.　**Legume**

   　　　4b. Fruit developed from more than one carpel.

   　　　　5a. Fruit splitting in one of four ways: along line of carpel union, along middle of each carpel, by pores, or along a circular, horizontal line.　**Capsule**

   　　　　5b. Fruit of two carpels with shared, central partition. Carpels split at maturity; seeds attached to partition.　**Silique**

3b. Indehiscent. Fruit remains closed at maturity.

    4a. Pericarp with a winglike extension. **Samara**

    4b. Pericarp without a winglike extension.

        5a. Pericarp a hard, woody shell. **Nut**

        5b. Pericarp not a hard, woody shell.

            6a. Seed attached to pericarp at one point. **Achene**

            6b. Seed coat entirely fused with pericarp. **Caryopsis** or grain

1b. *Aggregate fruit.* Fruit developed from several separate ovaries within a single flower.

1c. *Multiple fruit.* Fruit developed from fused parts of separate flowers within an inflorescence.

Common Name of Fruit	General Fruit Type	Ovary Position
_____	_____	_____
_____	_____	_____
_____	_____	_____
_____	_____	_____
_____	_____	_____
_____	_____	_____
_____	_____	_____
_____	_____	_____
_____	_____	_____
_____	_____	_____
_____	_____	_____
_____	_____	_____
_____	_____	_____

PART B   SEED DISPERSAL MECHANISMS

Fruits serve to disperse seeds. Some seeds may be dispersed by abiotic factors, such as wind or water. For example, trichomes that catch the wind or allow the fruit or seed to float. Other plants have evolved seed dispersal mechanisms that rely on animals—for example, fleshy fruits may be eaten by animals and the undigested seeds carried to new locations and deposited in dung. The mode of dispersal is usually obvious. What may not be as obvious is the source of the structure that promotes seed dispersal. In this portion of the exercise, various fruits and seeds representing different mechanisms of dispersal will be displayed. Try to determine the origin of the different dispersal structures.

MATERIALS

- prepared slide of cotton fruit cross section with mature seeds
- variety of fruits and seeds representing biotic and abiotic dispersal mechanisms
- compound microscope

PROCEDURE

1. Cotton Fibers

   Obtain a slide of a cotton fruit in cross section and examine the preparation with a compound microscope. The commercial fibers harvested from cotton plants come from epidermal hairs of the seed coat of cotton plants. These seed coat fibers have secondary walls and evolved to aid in dispersal by wind. Locate a cotton seed on the slide and focus on the seed coat hairs. The cotton fiber cells are numerous, covering the entire outer surface of the seed.

2. Fruit and Seed Dispersal

   As you examine the specimens, try to determine what plant parts are modified to aid seed dispersal.

   a. Biotic Dispersal

   Examine the **biotically dispersed** fruits and seeds on display. Record the species, whether the entire fruit or just seeds are dispersed, and the dispersal agent.

Species	Fruit or Seed	Biotic Agent of Dispersal
_____	_____	_____
_____	_____	_____
_____	_____	_____
_____	_____	_____
_____	_____	_____
_____	_____	_____
_____	_____	_____
_____	_____	_____

b. Abiotic Dispersal

Examine the fruit and seed **abiotic dispersal** display. Record the species, whether the entire fruit or just seeds are dispersed, and the dispersal mechanism.

Species	Fruit or Seed	Dispersal Mechanism
_____	_____	_____
_____	_____	_____
_____	_____	_____
_____	_____	_____

_____

QUESTIONS FOR THOUGHT AND REVIEW

1. What is a fruit? _____
_____
_____

2. Are fruits unique to angiosperms? Explain.
_____
_____
_____

3. List the parts of the pericarp._____
_____

4. Name two flower tissues that commonly become fruit accessory tissue._____
_____
_____

5. Speculate on the origin of seedless fruits.
_____
_____
_____
_____

6. What is the fruit type of an orange?_____
_____

Why is it classified as this fruit type?_____
_____

7. What is the origin of the edible part of an apple fruit?_____
_____

8. Describe an example of seed dispersal by an ant.
_____
_____

9. Describe an example of seed dispersal by water.
_____
_____

10. Describe an example of seed dispersal by a mammal. _____
_____
_____
_____

Name: _____

Section Number: _____

FRUIT MORPHOLOGY AND SEED DISPERSAL

1. What is the definition of a botanical fruit? _____

_____

2. List the three tissues of the pericarp.

    a. _____

    b. _____

    c. _____

3. From which floral tissue is the pericarp derived? _____

_____

4. Contrast simple, aggregrate, and multiple fruits. _____

_____

_____

_____

5. List two functions of fruits.

    a. _____

    b. _____

6. When referring to fruits, what is accessory tissue? _____

_____

_____

7. List two modes of biotic seed dispersal.

    a. _____

    b. _____

8. List two means of abiotic seed dispersal.

    a. _____

    b. _____

PLANT GROWTH AND DEVELOPMENT

OBJECTIVES

1. Identify plant growth associated with gravitropism and phototropism and identify the environmental signals that trigger these growth responses.

2. Understand and identify at least two patterns of plant growth responses associated with the pigment phytochrome.

3. Describe two controls of seed germination.

4. Describe the phenomena of root initation on cuttings and apical dominance and discuss the role of auxin in these events.

5. Describe a bioassay using gibberellic acid and dwarf pea plants.

6. Describe the roles of auxin and cytokinin in maintenance and differentiation of tissue cultures.

TERMINOLOGY

abscissic acid	gibberellin
apical dominance	gravitropism
auxin	hormone
bioassay	imbibition
callus	phototropism
cytokinin	phytochrome
ethylene	plant growth regulator
etiolation	red light
far-red light	scarification
germination	tissue culture

INTRODUCTION TO PLANT GROWTH AND DEVELOPMENT

Although plants lack the nerves and muscles of animals, they do exhibit remarkable adaptations that enable them to sense their environment and respond to environmental signals and changes. Typical reactions of plants include responses to gravity, light, and moisture. Anyone who has grown a house plant near a window is familiar with the consequence of not regularly turning the plant; plant stems and leaves tend to

grow toward light—particularly in the low-light conditions typical of most homes. In addition to sensing and responding to their environment, plants grow and develop under the influence of internal growth regulators. There are five well-documented classes of **plant growth regulators**, or plant **hormones: auxin, cytokinin, gibberellin, abscissic acid**, and **ethylene**. The response of a plant to these hormones depends on combinations of factors—for example, hormone concentration, specific plant tissue, age or developmental state of the plant, and environmental conditions. In this exercise, we will explore a few of the responses of plants to environmental signals and to specific hormones.

PART A PLANT RESPONSES TO THE ENVIRONMENT

Tropisms are plant growth curvatures in response to environmental stimuli such as gravity or light. **Gravitropism**, or curvature in response to the pull of gravity, is a common tropic response in plants. Young, primary roots typically exhibit positive gravitropism. The roots grow toward the earth. In primary roots, the root cap is the site of gravity perception. One of the most familiar plant curvature responses is **phototropism**, or curvature in response to a unidirectional source of light. Plants are sensitive to both the quantity and the quality of light in their environment. In order to **germinate**, certain species of seeds require exposure to light at the time of **imbibition**. For example, dark-grown seedlings typically exhibit spindly growth in response to light that is poor in **red light**. **Phytochrome** is a photoreversible pigment and controls these responses. Exposure to **far-red light** converts phytochrome to the P_r form. When a plant is exposed to red light, phytochrome converts to the biologically active form called P_{fr}. Biologically active phytochrome causes seed germination in phytochrome-sensitive seeds and causes spindly growth in seedlings exposed to light poor in red wavelengths. Some seeds require a **scarification** treatment to speed germination. In these plants, the seed coat must be broken for imbibition to occur. In nature, scarification might occur in seeds that are tumbled about against rocks in flowing water.

MATERIALS

- four sets of 2-day-old pea seedlings in beakers
- two *Coleus* plants with different light sources
- sets of lettuce seeds
- sets of etiolated seedlings
- sets of morning glory seeds

PROCEDURE

1. Gravitropism in a Primary Root

   Two days ago, pea seeds with primary roots emerging were placed in four beakers so that the roots were visible through the glass. Seeds with intact primary roots were placed so that the primary roots were either horizontal or vertical. Seeds with the tips of the roots removed were also placed either horizontally or vertically. The intact roots should exhibit downward growth—or positive gravitropism. The decapitated roots are missing root caps and will not exhibit gravitropism.

2. Gravitropism versus Phototropism in a Stem

   Your instructor will point out the location of this demonstration. Two days prior to lab, a *Coleus* plant labelled as plant A was placed on its side and exposed to light from below. A control plant (plant B) was placed upright and exposed to general illumination, not unidirectional light as for plant A. When compared with the control plant, it should be easy to see that the leaves of plant A have turned to expose their upper surfaces to the light source from below. This is a positive phototropic response. However, the stem tip is more responsive to gravity than light. Note that the stem tip of plant A is turned away from the earth and away from the light source. The stem tip is exhibiting negative gravitropism.

3. Light Control of Seed Germination

   Some seeds, such as Grand Rapids lettuce seeds, require light to germinate. Phytochrome controls this response. Examine the display of petri dishes containing lettuce seeds. There is a control dish, which was kept in the dark, and three experimental dishes. Each experimental dish was exposed to either red light, far-red light, or red light immediately followed by far-red light. Compare germination percentages among the four dishes of seeds. Seeds kept in the dark or exposed to far-red light should remain ungerminated. Seeds exposed to red light should germinate.

4. Etiolation in Seedlings

   Examine the display of dark-grown and light-grown seedlings. Note the multiple features that distinguish the **etiolated** (dark-grown) growth from the growth of seedlings kept in the light. Think about how the features of etiolated plants contribute to the plant's survival.

5. Scarification of Seed Coat and Germination

   Two sets of morning glory seeds are on display. One set is the control. Several days ago these seeds were placed on moist filter paper in a petri dish. In the experimental set the seed coats were abraded (the process of scarification) before placing the seeds on the moist filter paper. Note that the nonscarified seeds have not germinated. Determine the percentage of germination in the scarified set.

PART B PLANT RESPONSES TO HORMONES

Plant growth is also regulated by plant hormones. Auxin is involved in many plant growth processes and two processes, root initation in cuttings and **apical dominance**, are featured in this exercise. Gibberellins are another class of plant hormone and different concentrations of a gibberellin will be used in a **bioassay** experiment to demonstrate the effect of gibberellin on returning normal growth to dwarf pea plants. In addition, a portion of the exercise includes a demonstration of the interaction of auxin and cytokinin in maintaining tissue cultures and stimulating the formation of shoots or roots in **tissue cultures**.

MATERIALS

- sets of rooted cuttings
- sets of plants demonstrating apical dominance
- pots of dwarf pea plants
- different concentrations of gibberellin in flasks with individual pipettes
- centimeter scale
- tissue culture sets

PROCEDURE

1. **Auxin and Root Initiation**

   Auxin can stimulate the initiation of adventitious roots at the cut surfaces of cuttings. The demonstration shows the effect of dipping cuttings in auxin solutions of different concentrations. Note which concentration causes the most roots to form on the cutting. Also note the differences in the lengths of the adventitious roots. Auxin inhibits root elongation in established roots.

2. **Auxin and Apical Dominance**

   A plant with very little or no axillary bud growth exhibits apical dominance. If the shoot apex of such a plant is removed, the axillary buds begin to grow and expand. If auxin is applied to the stump left when the shoot apex was removed, the axillary buds remain inhibited. Examine the three plants in the apical dominance demonstration. One plant is the control, a second was decapitated and plain lanolin was placed on the stump, a third plant was decapitated and lanolin with auxin was placed on the stump. The control plant and the plant treated with auxin should continue to exhibit apical dominance. The decapitated plant not treated with auxin should have at least a few elongating axillary buds.

3. **Gibberellin and Growth of Dwarf Peas— A Bioassay**

   a. **Setup**

   Each lab section will be divided into five groups of students, and each group will be responsible for preparing two pots of dwarf pea plants. There should be five plants in each pot. Apply a strip of tape all the way around each pot. This will be used to record individual plant measurements for the plants around the periphery of the pots. Apply an additional strip near the bottom of each pot to record your group number, lab day, lab time, room number, and the measurements for the central plants. Record the data on the tape with a Sharpie® pen. (Do not use an ink pen, the writing will be lost when the plants are watered).

   Take the pots to your bench space and inspect one of the plants. Your first task is to find the shoot tip—not an easy job, because young leaves can be mistaken for the main plant axis. Look first at a leaf low on the plant. The leaf is compound: starting at the base of the petiole, it has a pair of bladelike stipules, three pairs of leaflets, and a pair of tendrils. Push aside the stipules to locate the axillary bud. Now move up the plant, tracing the stem to its tip, using axillary buds and leaflets as clues.

   With a centimeter scale, measure the height of the stem (in mm), from the top of the seed to the shoot tip. (Brush away the vermiculite to expose the seed.) Record this initial height of the plant on the strip of tape just below the plant and in Table 14-1. Continue measuring and recording heights for each plant in both pots.

   Five solutions of the plant hormone gibberellin (GA) will be available on the lab bench. Your instructor will assign one of the solutions to your group, and you will apply this solution to all of the plants in your pots. To apply the solution, use the pipette attached to the flask and carefully apply ONE drop to the shoot tip. If the drop runs off, do not apply another, but in your notebook state what happened. PLEASE RETURN THE PIPETTE TO THE APPROPRIATE FLASK. On the tape, record which solution you used.

   Similarly, treat the other plants. Record your treatment and measurements in Table 14-1, then return the pots to the instructor for a one-week incubation under the light bank.

   b. **Results**

   Locate the pots of dwarf pea plants that you treated with gibberellin last week. Measure each plant using the same technique as before, and record the results in Table 14-1. Subtract last week's height, and express the difference in millimeters. Record the data in Table 14-1.

Table 14-1   Growth of Dwarf Pea Plants in Response to GA Treatment; Goup Results

Group no.: _____

GA concentration:_____

	Pot 1		
Plant	Height (mm) after treatment = A	Initial height (mm) = B	Growth in height (mm) = A − B
1			
2			
3			
4			
5			

(continued)

Pot 2

Plant	Height (mm) after treatment = A	Initial height (mm) = B	Growth in height (mm) = A − B
1			
2			
3			
4			
5			

Average growth in height (total growth in height for all 10 plants divided by 10): _____

Standard deviation: _____

Calculate the average growth in height and the standard deviation, and record the calculated data in Table 14-2. When all lab data have been collected by students in the laboratory section, record the remaining lab data in Table 14-2.

Table 14-2   Growth of Dwarf Pea Plants in Response to GA Treatment; Lab Results

Group no.	[GA]	Average growth in height (mm)	Standard deviation
1			
2			
3			
4			
5			

c. Analysis

To find the gibberellin concentration in the unknown solution, first you must construct a graph (Figure 14-1). Show growth on the vertical axis and log [GA] on the horizontal axis, from lowest to highest concentration. (If [GA] = 10^{-6} M, then log [GA] = −6.) Draw a straight line that appears to be the best compromise between the data points. Next, mark the graph's vertical axis to show the growth of the plant treated with the unknown solution. Draw a horizontal line from this point. Where this line crosses the known concentration line, drop a vertical line to the horizontal axis. This is your estimate of the unknown GA concentration.

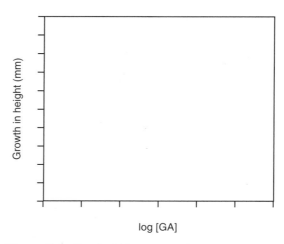

Figure 14-1   Graph of bioassay results.

4. Tissue Cultures and Organ Initiation

Examine the tissue culture demonstrations on display. Describe the texture of the **callus** (undifferentiated tissue). Is it smooth, granular, bumpy? In these cultures, note which hormone treatments induced shoot growth, root growth, or growth of both shoots and roots from callus tissue. Speculate as to how these treatments inform us about the roles of hormones in intact plants or in other plant species.

QUESTIONS FOR THOUGHT AND REVIEW

1. What external signal controls the growth direction in roots? _____

_____

_____

2. How does growth orientation relate to root function? _____

_____

_____

3. In the gravitropism versus phototropism demonstration, what parts of the experimental plant appear to sense the environment and how did these parts respond? _____

_____

_____

_____

4. What light-detecting pigment is involved in etiolated seedlings? _____

_____

5. Write an equation describing the response of phytochrome to red light and to far-red light.

_____

_____

6. How do features of etiolated seedlings contribute to their survival? _____

_____

_____

_____

7. What are auxin and gibberellin? _____

_____

_____

_____

_____

8. How would you attempt to induce roots on cut stems? _____

_____

_____

_____

_____

9. Explain how you could use the results of a bio-assay to estimate the hormone concentration in the extract that was assayed. _____

_____

_____

_____

_____

10. What hormone or hormones would you use to keep tissue as callus? _____

_____

_____

_____

Name:_____

Section Number: _____

PLANT GROWTH AND DEVELOPMENT

1. List the five common plant hormones.

    a. _____

    b. _____

    c. _____

    d. _____

    e. _____

2. List two plant responses to stimuli from the environment.

    a. _____

    b. _____

3. What specific environmental factor causes phototropism in plants? _____

    _____

4. Describe scarification. _____

    _____

    _____

5. List two plant growth responses involving auxin.

    a. _____

    b. _____

6. When might you use a bioassay in the study of plant hormones and plant response?_____

    _____

    _____

7. Describe a practical gardening use of apical dominance. _____

    _____

    _____

8. What is callus? _____

    _____

    _____

EXERCISE **15**

BACTERIA AND CYANOBACTERIA

OBJECTIVES

1. Differentiate prokaryotic from eukaryotic cells.

2. Describe the use of specific stains and differential stains in bacteriology.

3. Use the Gram stain technique.

4. Describe and use methods requiring sterile technique in handling bacteria.

5. Describe a culture technique that differentiates between gram-positive and gram-negative bacteria and between lactose-fermenting and nonlactose-fermenting bacteria.

6. If given a suitable photograph or specimen, identify bacterial endospores.

7. Define antibiotic and describe a procedure to test for effectivness of an antibiotic on bacterial growth.

8. Define nitrogen fixation and explain the roles of *Rhizobium* and leguminous plants in a mutualistic association involving root nodules.

9. If given an appropriate photograph or specimen, identify cyanobacteria.

10. Describe the difference among unicellular, colonial, and filamentous cyanobacteria.

11. Understand the function of heterocysts and akinetes in cyanobacteria.

12. Explain the roles of *Anabaena* and *Azolla* in a nitrogen-fixation mutualistic association.

TERMINOLOGY

aerobic	filament
agar	Gram stain
akinete	heterocyst
antibiotic	heterotroph
archaebacterium	legume
autotroph	mutualism
bacillus	nitrogen fixation
capsule	peptidoglycan
chlorophyll a	phototroph
coccus	phycobilin
colony	prokaryote
differential growth	root nodule
medium	spirillum
differential stain	symbiosis
endospore	unicell
eubacterium	

GENERAL INTRODUCTION TO BACTERIA

Bacteria are one of the most important groups of organisms on Earth. They were the first living organisms to evolve, appearing in the fossil record about 3.5 billion years ago. They outnumber all other organisms, they play vital ecological roles, and they significantly influence human welfare. Compared with eukaryotes, bacteria, which are **prokaryotes**, have simple structures; however, they exhibit a large diversity in metabolism. Recent ribosomal RNA sequence data support classifying prokaryotes in two major lineages: **eubacteria** and **archaebacteria**. Archaebacteria evolved metabolic pathways that enable them to thrive in harsh environments, such as hot, acid pools found near sites of geothermal activity or regions of very high salinity. Most bacteria are **heterotrophs**, but some are **autotrophs**. All **phototrophic** bacteria obtain energy from sunlight, but use different substrates and pigments during photosynthesis. Heterotrophic and phototrophic eubacteria are the subjects of this laboratory exercise.

PART A IDENTIFICATION OF BACTERIA

1. Bacterial Cell Shapes

    Cell shape is one of the characteristics used to identifiy specific groups of bacteria, and the three most common cell shapes are **coccus**, **bacillus**, and **spirillum**. Cocci are spherical cells; bacilli are straight, rod-shaped cells; and spirilli are long, curved cells. Refer to Figure 15-1.

MATERIALS

- prepared slide of three forms of mixed bacteria, Gram-stained
- compound microscope with oil immersion lens
- immersion oil

PROCEDURE

Using a compound microscope at high power, locate **colonies** of bacteria. Apply a drop of immersion oil to the slide and slowly rotate the oil immersion lens into the oil. Observe the different shapes of bacteria on the prepared slide. Identify cocci, bacilli, and spirilli. Note the dark purple cells that indicate gram-positive bacteria and the pink cells indicating gram-negative bacteria.

2. Bacterial Structure Stains

    Living preparations of bacteria can be observed with the light microscope. However, fixation and staining increases cell visibility and preserves the preparation for further use. Preservation of the general shape of a bacterial cell is accomplished by heat fixation. To heat fix bacteria, bacteria are smeared across the surface of a microscope slide and the slide is gently heated over a flame, leaving an attached, dried layer of killed bacteria on the slide. (Chemical fixation must be used to preserve internal structures within the cell.) Along with fixation, differential or specific stains can be used to reveal structures such as a **capsule** surrounding a bacterial cell or resistant, dormant cells called **endospores**.

MATERIALS

- prepared slide of *Klebsiella pneumoniae* with capsule stain
- prepared slide of *Clostridium tetani* with spore stain
- compound microscope with oil immersion lens
- immersion oil

PROCEDURE

a. *Klebsiella pneumoniae*

    Use a compound microscope and oil immersion to examine the capsule surrounding *Klebsiella pneumoniae* cells. Bacterial capsules are composed of organized layers of polysaccharides and help protect the cells from desiccation and from microbial attack. *K. pneumoniae* is a member of the coliform group of bacteria. Presence of coliform bacteria in water supplies is an indication of polluted water.

b. *Clostridium tetani*

    Use a compound microscope and oil immersion to examine the *Clostridium tetani* slide. This organism is a common anaerobic, soil-inhabiting bacterium, and it is the cause of tetanus. Look for round, terminal endospores. Endospores are capable of surviving for long periods of time under less than favorable conditions.

3. Differential Staining: The Gram Stain

    The **Gram stain** is the most commonly used staining technique for bacterial identification. It is a **differential stain** because, based on cell wall composition, it divides bacteria into two taxonomic groups. The cell walls of gram-positive bacteria consist of a layer of **peptidoglycans** with polysaccharides. Gram-negative bacteria cell walls are more complex; they consist of two layers, an outer layer composed of lipoproteins and lipopolysaccharides (similar in structure to the cell

Figure 15-1   Mixture of stained bacterial cultures as seen in the light microscope. (**a**) Coccus ◯, (**b**) bacillus ⬭, and (**c**) spirillum ⌇.

membrane) and an inner layer composed of peptidoglycans. During Gram staining, the walls of gram-positive bacteria combine with the primary dye (crystal violet) and appear dark purple; gram-negative bacteria do not combine with crystal violet.

Since all smears must be air-dried before proceeding with the staining procedure, it will save time to make all three smears first. However, read through the entire procedure before beginning. Follow the procedure for Gram staining, and classify the three bacteria species (*Bacillus cereus*, *Pseudomonas pudica*, and *Escherichia coli*) as gram-positive or as gram-negative.

MATERIALS

- culture of *Bacillus cereus* grown on nutrient agar
- culture of *Pseudomonas pudica* grown on nutrient agar
- culture of *Escherichia coli* grown on nutrient agar
- inoculation loop
- Bunsen burner or alcohol lamp
- jar of sterile distilled water
- crystal violet stain in a dropper bottle
- Gram's iodine stain in a dropper bottle
- safranin stain in a dropper bottle
- 95% ethyl alcohol destain in a dropper bottle
- prepared slides of gram-stained bacteria
- clean microscope slides
- immersion oil
- compound microscope with an oil immersion lens

PROCEDURE

Smear preparation

a. Heat an inoculation loop in the flame of a Bunsen burner or alcohol lamp until the loop just turns red. This will sterilize the loop.

b. Allow the loop to cool briefly (do not wave it around in the air), or quickly cool the loop by plunging it into sterile **agar** at the edge of a petri dish.

c. Pick up a small amount of bacteria from a colony and mix with a small drop of water on a clean microscope slide. (Smears that are too thick can give false gram-positive results; a small drop of water will speed the drying process.)

d. Spread the bacteria-water drop to form an even film.

e. Allow the slide to air dry. Sterilize the inoculation loop.

Stain technique

f. Heat fix the bacteria to the slide by quickly passing the slide, smear side up, through the flame of a Bunsen burner two to three times. Do not hold the slide in the flame.

g. Stain with crystal violet for 30 seconds. Holding the slide at an angle in a sink, gently wash the slide in tap water for about 2 seconds, then drain excess water from the slide surface. Air dry.

h. Apply iodine solution to the bacteria smear for 1 minute. Rinse the slide with tap water and air dry it.

i. Holding the slide at a 45° angle in a sink, apply destain alcohol drop by drop until no more crystal violet runs from the smear. Quickly rinse with water. (Be careful not to over- or under-destain; the destain works rapidly, 5 to 10 seconds is usually long enough.)

j. Apply safranin counterstain to the smear for 30 seconds. Rinse the slide with tap water and air dry it.

Examining stained preparation

k. The smear is now ready for examination. To examine the preparation with the oil immersion objective lens of a compound microscope, a drop of immersion oil can be placed directly on the smear without using a coverslip.

l. If the smears were stained successfully, gram-positive cells will retain the dark purple of the crystal violet. Gram-negative bacteria will appear pink because they lose crystal violet during the destaining procedure and stain pink with the safranin counterstain. Record the results in Table 15-1.

Table 15-1 Gram Stain Results

Organism	Gram + or −
Bacillus cereus	
Pseudomonas pudica	
Escherichia coli	

4. Sterile Technique and a Differential Growth Medium

In this exercise sterile technique will be used to inoculate agar with bacteria, and a **differential growth medium**, MacConkey agar, will be employed to distinguish between lactose-fermenting and nonlactose-fermenting bacteria. MacConkey agar contains lactose as a carbon source, the pH indicator neutral red, and bile salts that inhibit growth of gram-positive bacteria. When lactose is fermented, there is a localized pH drop, and lactose-fermenting bacteria absorb neutral red, turning the colonies of bacteria bright pink. Colonies of bacteria that cannot ferment lactose remain translucent.

MATERIALS

- sterile cultures of *Bacillus cereus* grown on nutrient agar
- sterile cultures of *Pseudomonas pudica* grown on nutrient agar
- sterile cultures of *Escherichia coli* grown on nutrient agar
- inoculation loop
- Bunsen burner or alcohol lamp
- jar of sterile distilled water
- petri dishes with nutrient agar
- petri dishes with MacConkey agar
- clean microscope slides and coverslips
- immersion oil
- compound microscope with an oil immersion lens

PROCEDURE

Sterile technique

To start, vigorously wash your hands, and wash the lab bench working surface with disinfectant. Also, after handling microorganisms, always wash your hands *before* leaving the laboratory.

a. Sterilize an inoculation loop in the flame of a Bunsen burner or alcohol lamp; cool the loop.

b. With the sterile loop, carefully pick up a very small amount of bacteria. (You may dip the sterile loop into a jar of sterile water, wetting the loop, before picking up the bacteria inoculum.)

c. Hold the inoculated loop near the edge of the petri dish to be inoculated.

d. To transfer the bacteria on the loop to the plate, carefully streak the loop in a loose zigzag pattern across one-fourth of the agar surface. Keep the loop on the agar surface, do not jab it into the agar, and do not go back over any areas already streaked; you want to spread the bacteria over the surface.

e. Sterilize the loop in the flame again and allow it to cool.

f. Turn the plate 90° counterclockwise, and touch the loop into the lower right region previously streaked. Now streak the loop over a fresh region of the agar, covering about one-fourth of the agar surface.

g. Sterilize the loop and repeat step (f), beginning at the edge of the region last streaked.

h. Sterilize the loop to redness to kill any remaining bacteria.

Inoculating McConkey differential growth medium

i. Use fresh, sterile bacteria cultures. Working in groups of three, each student will choose one of three species of bacteria (*Bacillus cereus*, *Pseudomonas pudica*, or *Escherichia coli*) and, using sterile technique, streak two agar plates—one plate with nutrient agar and one plate containing MacConkey agar. Therefore, within each group there will be a total of six plates, and each organism will be streaked onto both a nutrient agar plate and a MacConkey agar plate. The plates will be incubated at room temperature, and the results evaluated during the next laboratory period.

Examining results

j. Note the growth and color of the three bacterial organisms streaked during the previous laboratory period. Compare the growth of each species on nutrient agar with growth on MacConkey agar (remember, gram-positive bacteria do not grow well on MacConkey agar). In Table 15-2, indicate the relative growth of the bacteria on the MacConkey agar, and indicate whether or not the organisms are lactose fermenters.

Table 15-2 MacConkey Differential Medium Results

Organism	Relative growth on MacConkey agar	Lactose fermenter + or −
Bacillus cereus		
Pseudomonas pudica		
Escherichia coli		

PART B BACTERIAL SENSITIVITY TO ANTIBIOTICS

Antibiotics are natural, organic substances that kill or inhibit the growth of microorganisms. Many bacteria species produce antibiotics, and the response of other bacteria to these growth inhibitors varies with the species. Generally, antibiotics inhibit the synthesis of bacterial cell walls. Narrow-spectrum antibiotics inhibit growth in a few species of microorganisms, and broad-spectrum antibiotics are effective against many different kinds of microorganisms. Bacteria also vary in their sensitivity to antibiotic concentration.

Different techniques are used to test the level of antibiotic activity against specific species of bacteria. For example, the disk diffusion antibiotic test is easy and inexpensive. In this test, disks soaked with different antibiotics are placed on agar that has been inoculated with one bacterium species. Because the antibiotics diffuse from the disks into the agar, antibiotic concentrations are higher near the disks and lower away from the disks. If the bacterium is sensitive to an antibiotic, a clear ring is present around the disk. The larger the clear area around a specific antibiotic disk, the more susceptible the bacterium is to that antibiotic.

MATERIALS

- two different prepared bacterial sensitivity plates

PROCEDURE

1. Use Table 15-3 to record the species of bacteria used to inoculate each plate.

2. Record the kinds of the antibiotic disks used on each plate.

3. Record the extent of growth around each antibiotic disk. Use the following notation to indicate extent of growth:

   −   growth not inhibited

   +   growth slightly inhibited

  + +   growth moderately inhibited

 + + + growth greatly inhibited

Table 15-3 Antibiotic Sensitivity

	Extent of inhibition	
Antibiotic	Bacterium species _____	Bacterium species _____
A		
B		
C		
D		
E		

PART C BACTERIAL NITROGEN FIXATION AND MUTUALISM

The earth's reservoir of nitrogen is in the form of atmospheric nitrogen gas, N_2. Nitrogen is a major plant nutrient, but the N_2 form cannot be used by plants. However, certain bacteria, such as *Rhizobium*, are capable of fixing atmospheric gaseous nitrogen into a form that is usable by plants. The reduction of N_2 to ammonia is called **nitrogen fixation**.

Members of the **legume** family of plants can form beneficial **symbiotic** associations with nitrogen-fixing *Rhizobium*. A beneficial symbiotic association is called **mutualism**. The bacteria invade root hair cells, stimulating the root to grow small nodules. Within the **root nodules**, *Rhizobium* fixes atmospheric nitrogen, the plant absorbs the fixed nitrogen, and in exchange the bacterium receives protection and food from the plant.

MATERIALS

- bean plants inoculated with *Rhizobium*
- prepared slide of bean root with attached nodule
- compound microscope

PROCEDURE

1. Bean plants with *Rhizobium* root nodules

   Examine the root system of a living *Rhizobium*-inoculated bean plant and look for root nodules along the roots.

2. *Rhizobium* root nodule slide

   Use a compound microscope to examine a prepared slide of a bean root with a *Rhizobium* root nodule. Differentiate among root tissue, root nodule tissue, and bacteria.

GENERAL INTRODUCTION TO CYANOBACTERIA

The cyanobacteria are photosynthetic eubacteria; but, like plants and algae, cyanobacteria use **chlorophyll a** as their primary photosynthetic pigment. Although cyanobacteria do not possess chloroplasts, they do have thylakoids that are not enclosed within a membrane. In fact, cyanobacteria probably gave rise to chloroplasts through endosymbiosis. In addition to chlorophyll a, cyanobacteria have accessory photosynthetic pigments called **phycobilins**. Phycobilins are linked to proteins forming phycobiliproteins that are located on the surface of the thylakoid membranes. These ancient prokaryotes (also known as blue-green algae) are important primary producers in marine and freshwater ecosystems, are common in terrestrial environments, and are partners with fungi, plants, and animals in numerous symbiotic associations. In some of these mutualistic associations, the cyanobacterium contributes fixed nitrogen. In **aerobic** conditions, some cyanobacteria differentiate specialized cells called **heterocysts** that exclude oxygen and permit the fixation of atmospheric nitrogen into ammonia. Only in the absence of oxygen can nitrogenase catalyze the conversion of N_2 to ammonia.

PART A IDENTIFICATION OF CYANOBACTERIA

Cyanobacteria morphology varies from **unicells**, to aggregations or colonies of cells, to unbranched or branched **filaments**. Most cells are surrounded by a mucilaginous sheath, and some filaments are capable of a gliding movement. Certain cyanobacteria develop specialized cell types called heterocysts and **akinetes**. Heterocysts are sites of nitrogen fixation and akinetes are thick-walled resistant cells that can survive dormancy.

MATERIALS

- culture or prepared slide of *Gloeocapsa*
- culture or prepared slide of *Merismopedia*
- culture or prepared slide of *Oscillatoria*
- culture or prepared slide of *Gloeotrichia*
- culture or prepared slide of *Scytonema*
- culture or prepared slide of *Stigonema*
- microscope slides
- coverslips
- pipettes with bulbs
- dissecting needles
- compound microscope

PROCEDURE

1. *Gloeocapsa*

   Make a wet mount from the living culture of *Gloeocapsa* and observe. In this genus, individual cells form aggregations within a distinctive mucilage.

2. *Merismopedia*

   Make a wet mount of *Merismopedia* and observe the colonies of cells. These colonies develop in a single plane of rows and columns of cells within mucilage.

3. *Oscillatoria*

   From the living culture of *Oscillatoria*, take a small sample of filaments and place the sample on a microscope slide. Tease apart the filaments, which are unbranched, add a coverslip, and observe with a compound microscope. Focus on a single filament, noting the shape of individual cells. *Oscillatoria* cells within the filament are disc-shaped, and the filaments exhibit a subtle gliding movement.

4. *Gloeotrichia*

   Gloeotrichia is capable of nitrogen fixation and the unbranched, tapering filaments demonstrate some division of labor between cells. If a living culture of *Gleotrichia* is available, transfer a piece of a small gelatinous globule to a microscope slide. Add a coverslip and observe with the compound microscope. Note that the globule consists of many filaments of *Gloeotrichia*. Individual *Gloeotrichia* filaments have terminal heterocysts, subterminal akinetes, and photosynthetic cells. All but the heterocyst is surrounded by a gelatinous sheath.

5. *Scytonema*

   These organisms are unbranched filaments, but superficially appear branched. This type of growth is called false-branching. Make a wet mount from the living culture and observe with the compound microscope.

6. *Stigonema*

   This genus exhibits true branching. Make a wet mount and compare with the false-branching of *Scytonema*.

PART B CYANOBACTERIAL NITROGEN FIXATION AND MUTUALISM

The cyanobacterium *Anabaena* forms a mutualistic association with a water fern, *Azolla. Azolla* leaves shelter *Anabaena* and provide the cyanobacterial cells with carbohydrates. In exchange, *Anabaena* provides the plant with fixed nitrogen. *Azolla* is commonly grown in rice paddies, and any fixed nitrogen that leaks from the fern provides a natural source of nitrogen for the rice plants in the paddy.

MATERIALS

- prepared slide of *Anabaena*
- *Azolla* leaves with *Anabaena*
- microscope slides
- coverslips
- dH$_2$O in dropper bottle
- dissecting needles
- single-edge razor blade
- compound microscope

PROCEDURE

1. *Anabaena*

   To familiarize yourself with what you will be looking for, examine the *Anabaena* filaments on a prepared slide. Because of a constriction between adjacent cells, *Anabaena* filaments look like a string of beads. Note the size and shape differences between heterocysts and vegetative cells. Refer to Figure 15-2.

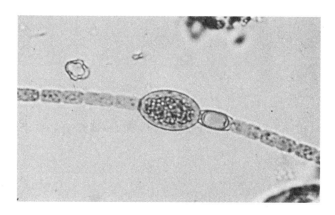

Figure 15-2 *Anabaena.* Simple filaments with heterocysts.

2. *Azolla* leaves with symbiotic *Anabaena*

   To see *Anabaena* inhabiting fern leaves, mince a portion of an *Azolla* leaf. To do this, place a piece of fern leaf in a drop of culture solution on a microscope slide and mince the leaf with a new razor blade. Add a coverslip and observe with the high power objective lens. Look for filaments with the small diameter and blue-green color characteristic of *Anabaena*. Also, look for heterocysts.

QUESTIONS FOR THOUGHT AND REVIEW

1. Sketch the shapes of the three bacterial cells that you stained with the Gram stain, and indicate whether each is gram-positive or gram-negative.

2. Name one function of a bacterial capsule._____

3. What is a bacterial endospore? _____

_____

_____

4. What does the Gram stain differentiate between?

_____

_____

5. What was the purpose of growing different species of bacteria on MacConkey agar?

_____

_____

_____

6. How might antibiotics be useful to bacteria in the wild?_____

_____

7. Name the bacterium responsible for nitrogen fixation in legumes. _____

8. What morphologies are present in the cyanobacteria?_____

_____

_____

9. What cell type is the site of nitrogen fixation in cyanobacteria? _____

10. How might *Azolla* benefit from association with *Anabaena*? _____

_____

_____

How might *Anabaena* benefit from association with *Azolla*? _____

_____

_____

_____

EXERCISE 15
LABORATORY QUIZ

Name:_____

Section Number: _____

BACTERIA AND CYANOBACTERIA

1. What defines a prokaryotic cell? _____

_____

2. List the three common bacterial cell shapes.

    a. _____

    b._____

    c. _____

3. What stain would you use to reveal a capsule surrounding a bacterial cell? _____

_____

4. In what habitat would you expect to find the tetanus-causing bacterium?_____

_____

5. Name the most common staining technique used to identify bacteria. _____

_____

6. In the laboratory exercise, why was sterile technique employed in preparation of the differential growth medium experiment?_____

_____

_____

_____

7. What is meant by nitrogen fixation?_____

_____

8. What photosynthetic pigment do cyanobacteria and plants have in common? _____

_____

EXERCISE 16

FUNGI AND FUNGUSLIKE PROTISTS

OBJECTIVES

1. In a sexually reproducing zygomycete, ascomycete, or basidiomycete fungus, identify vegetative hyphae and reproductive structures.

2. Given a mushroom, identify the stipe, cap, and gills or pores.

3. Describe the condition of n+n and identify the ploidy of any stage in the life cycle of a zygomycete, ascomycete, or basidiomycete fungus.

4. In an asexually reproducing zygomycete or ascomycete fungus, identify vegetative hyphae and reproductive structures.

5. Describe the active spore dispersal mechanism of the zygomycete fungus, *Pilobolus*.

6. Describe at least one way in which zygomycete, ascomycete, and basidiomycete fungi affect humans.

7. Briefly describe characteristics of four divisions of heteromorphic members of the kingdom Protista.

TERMINOLOGY

aeciospore	mycelium
ascocarp	mycorrhiza
ascospore	parasite
ascus	plasmogamy
basidiocarp	progametangium
basidiospore	rhizoid
basidium	saprophyte
cap	septate
chitin	sporangiophore
coenocytic	sporangium
conidiophore	spore
conidium	stipe
dikaryon	stolon
dikaryotic	suspensor
gametangium	teliospore
gill	uredospore
heterotroph	yeast
hypha	zygosporangium
karyogamy	

INTRODUCTION TO THE FUNGI

Fungi are not related to plants, but are distinct eukaryotic organisms placed in their own kingdom. However, fungi historically have been studied with other botanical organisms. Fungi are **heterotrophic** organisms, **saprophytic** or **parasitic**; and, although some appear green, none contain photosynthetic pigments. Although a few fungi, such as **yeasts**, are unicellular, most fungi have a multicellular body, the **mycelium**, composed of filaments called **hyphae**. The major cell wall polysaccharide in fungi is **chitin**—the same substance found in insect and crustacean exoskeletens. True fungi lack motile cells. Most fungi are terrestrial organisms and, along with bacteria, are ecologically important as decomposers. Fungi feed by secreting enzymes that break down complex organic matter into simple organic compounds and inorganic molecules. This process also greatly contributes to the environmental cycling of such molecules as carbon, nitrogen, and phosphorous. Fungi are both harmful and beneficial to plants. They are the leading cause of plant diseases, and they form complex mutualistic relationships, called **mycorrhizae**, with the roots of many vascular plants.

PART A SEXUAL REPRODUCTION AND THE DIVISIONS OF FUNGI

The taxonomic scheme used to separate fungi into divisions is based mainly on characteristics of sexual reproduction. There are four divisions of fungi. Three divisions with known sexual reproduction, the Zygomycota, the Ascomycota, and the Basidiomycota, and one division, the Deuteromycota, lacking known sexual reproduction.

1. Division Zygomycota—The Bread Molds

The members of this smallest division of fungi primarily live on decomposing animal and plant matter, but a few are parasitic on animals and plants. Zygomycetes have **coenocytic** hyphae with haploid nuclei. The hyphae arch along the substrate forming horizontal hyphae called **stolons**. Where stolons touch the substrate,

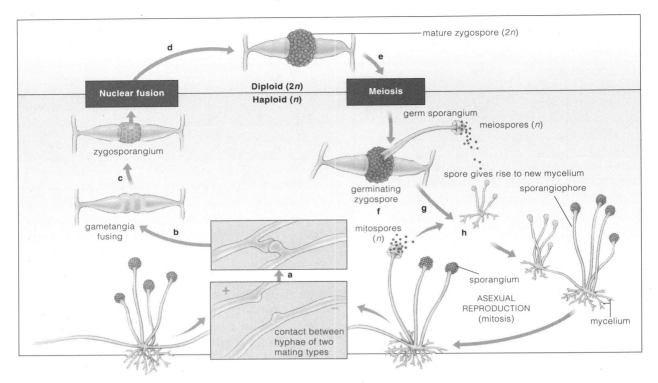

Figure 16-1 *Rhizopus* life cycle.

hyphae branch into the substrate and form **rhizoids**. During sexual reproduction, haploid hyphae of different mating types come together to produce a **zygosporangium**, also called a zygospore, where meiosis occurs. The short segments of hyphae that grow toward one another are called **progametangia**. Septation in each progametangium creates **gametangia**, which then fuse to form a zygosporangium. Upon germination, a germ sporangium emerges and produces haploid spores that start the next generation. Refer to Figure 16-1 as you examine the zygomycetes.

MATERIALS

- culture of *Gilbertella* with zygospores
- prepared slide of *Rhizopus stolonifer* zygosporangia, whole mount
- microscope slides
- coverslips
- dH₂O in dropper bottle
- dissecting needles
- dissecting microscope
- compound microscope

PROCEDURE

a. Culture of *Gilbertella* zygosporangia

Pick up a culture of *Gilbertella* zygosporangia and, using a dissecting microscope, examine the mycelium with the petri dish lid in place. A dark line marks where two different strains of mycelia have come together, forming golden-brown fuzzy zygosporangia. Open the petri dish, lift a tiny patch of mycelium from the surface, mount the mycelium in a small drop of water, and observe the preparation under a compound microscope. The zygosporangia are distinguished by their sculptured walls, **suspensors**, and color.

b. Prepared slide of *Rhizopus* zygosporangia

Use a compound microscope to examine the prepared slide of *Rhizopus* zygosporangia. Scan the slide in search of sexual structures at different stages of development, and look for the remnants of the suspensors that stand out from both sides of the zygosporangia. Mature zygosporangia will appear dark brown or black (Figure 16-2).

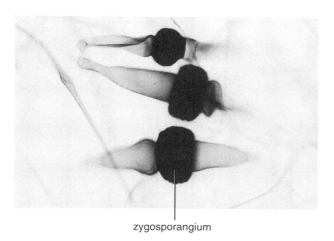

zygosporangium

Figure 16-2 *Rhizopus* zygosporangia.

2. Division Ascomycota—The Sac Fungi

Except for the unicellular yeasts, all ascomycetes are **septate**, filamentous fungi. Ascomycetes are economically important. Some grow as molds on food causing spoilage, some cause plant diseases such as powdery mildew, chestnut blight, and Dutch elm disease, but others are beneficial. For example, the edible morels and truffles are ascomycetes, and certain yeasts are important in wine-making, brewing, and baking. The defining feature of the ascomycetes is the formation of sexual spores inside a saclike cell called an **ascus** (Figure 16-3). An ascus forms when two haploid nuclei within a **dikaryotic** cell fuse to form a zygote nucleus. (A dikaryotic cell, or **dikaryon**, has two separate nuclei, the result of delay between **plasmogamy** and **karyogamy**.) The zygote nucleus undergoes meiosis, producing four **spore** nuclei. This event is followed by a mitotic division that produces eight spore nuclei. Cell walls form around each nucleus, and the mature ascus now contains eight **ascospores**. Upon ejection from the ascus, each ascospore germinates, producing the next generation of mycelia. In the majority of ascomycetes, clusters of asci form inside macroscopic structures called **ascocarps**.

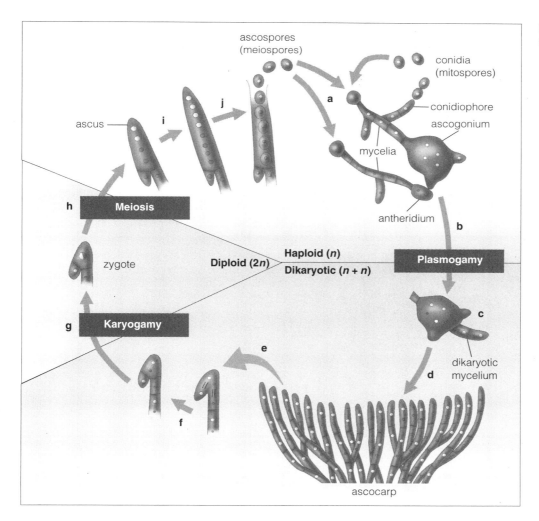

Figure 16-3 The life cycle of an ascomycete.

MATERIALS

- culture of *Ascobolus* with ascocarps
- prepared slide of a *Peziza* ascocarp
- variety of living and preserved ascocarps
- culture of sexual stage of *Saccharomyces*
- microscope slides
- coverslips
- dH$_2$O in dropper bottle
- dissecting needles
- single-edge razor blade
- dissecting microscope
- compound microscope

PROCEDURE

a. Culture of *Ascobolus* ascocarps

Use a dissecting microscope to inspect the culture of *Ascobolus*. The golden spheres on the agar surface are ascocarps. The spores are brown and may be visible on the petri dish lid and on the agar surface around the ascocarps. *Ascobolus* spores are shot into the air after the tips of the pressurized asci rupture. With a sharp razor blade, cut and remove a thin vertical section through an ascocarp. Try to keep the section thinner than 1 mm. Lay the ascocarp section in a drop of water on a slide, add a coverslip, and observe with a compound microscope. The asci are lined up side by side, each containing eight ascospores (Figure 16-4). Look for sterile as well as fertile hyphae, and watch for asci discharging spores.

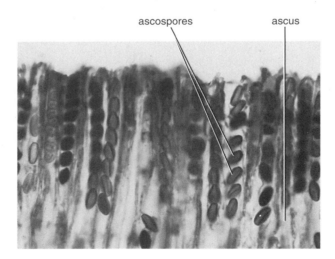

ascospores   ascus

Figure 16-4 *Ascobolus* asci, longitudinal section. Mature asci contain eight ascospores.

b. Prepared slide of *Peziza* ascocarps

Using the low power of a compound microscope, inspect a prepared slide of a *Peziza* ascocarp at low magnification and note the shape of the ascocarp. Then examine the slide with the 40× objective. The asci and ascospores should appear similar to those observed in the living *Ascobolus* ascocarps.

c. Ascocarp diversity

Look at the displayed specimens demonstrating different ascocarp morphologies. One type of ascocarp is open and dishlike (called an apothecium), a second type is partially closed and flasklike (a perithecium), and a third type is completely closed (a cleistothecium).

d. Yeasts

Yeasts are microscopic ascomycetes that do not form ascocarps. The yeast *Saccharomyces* reproduces sexually by making a naked ascus containing ascospores. Make a wet mount of the *Saccharomyces* culture and locate sexual structures.

3. Division Basidiomycota—The Club Fungi

Basidiomycetes include the familiar mushrooms and puffballs as well as serious agricultural pests called rusts and smuts. All basidiomycetes produce sexual spores, **basidiospores**, on a clublike structure, the **basidium** (Figure 16-5). Mushrooms are one example of macroscopic basidia-containing structures called **basidiocarps**. Typical basidiocarps consist of a **stipe**, a **cap**, and **gills** or pores. Each mass of mushroom-forming mycelium is dikaryotic. Therefore, the stalk, cap, and gills of a mushroom are composed of dikaryotic hyphae. Basidia are produced at the surface of the gills or pores. Although mushroom-forming basidiomycetes do not reproduce asexually, rusts and smuts make asexual **conidia**.

MATERIALS

- culture of *Coprinus* with basidiocarps at different stages
- prepared slide of *Coprinus* basidiocarp
- variety of living or preserved basidiocarps
- microscope slides
- coverslips
- compound microscope

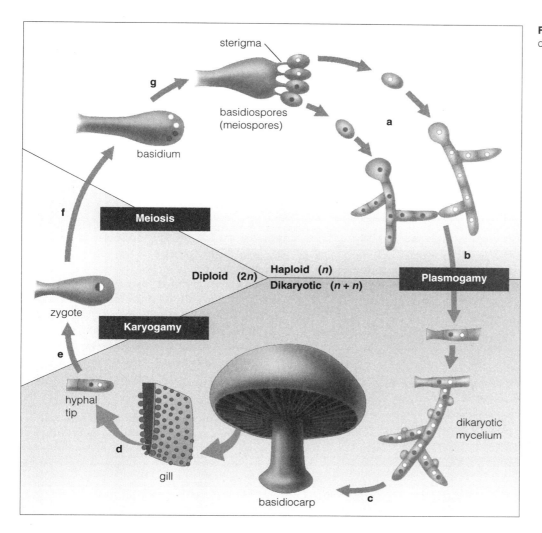

Figure 16-5 The life cycle of a basidomycete.

sterigma

g

basidiospores (meiospores)

basidium

a

Meiosis

f

Diploid (2n) **Haploid (n)**
Dikaryotic (n + n)

b

Plasmogamy

zygote

Karyogamy

e

hyphal tip

d

gill

dikaryotic mycelium

basidiocarp

c

PROCEDURE

a. Basidiocarp morphology

Agaricus bisporus is the common edible mushroom available in markets. Examine an *Agaricus* mushroom, identifying the stipe, cap, and gills.

b. Basidiocarp development

Culture of *Coprinus* basidiocarps

The culture exhibits different stages of *Coprinus* basidiocarp development, including vegetative mycelium, initiation and growth of basidiocarps, and spore discharge. Inspect the culture and look for these stages.

Prepared slide of a *Coprinus* basidiocarp

Examine a slide of *Coprinus* with a compound microscope. The sections through the gills show good examples of basidia, some with basidiospores still attached (Figure 16-6).

c. Basidiocarp diversity

In addition to the *Agaricus* mushroom, a variety of other basidiocarps is displayed. Compare gilled basidiocarps with those with pores, and compare mushrooms with puffballs. If you like, make a fresh mount of one of the

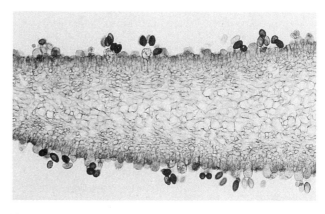

Figure 16-6 *Coprinus* gill, cross section, with basidia and basidiospores.

living, market mushrooms from the basidio-carp morphology station. To do so, take a 2×2 mm portion of a gill, place it in a drop of water on a slide, add a coverslip, and observe. Most of the tissue will be too thick, but the margins may show basidia with attached basidiospores. Use the 40× objective to see the basidia and spores.

PART B ASEXUAL REPRODUCTION IN FUNGI

Asexual spores are easy to make and energetically cheap to produce. These spores are produced in huge quantities, and some, such as *Alternaria*, are included in pollen counts because they cause human allergies. There are two basic means of asexual spore production: by **sporangia** and by conidia.

MATERIALS

- culture of *Rhizopus stolonifer* with sporangia
- prepared slide of *Rhizopus stolonifer* sporangia, whole mount
- demonstration of *Pilobolus* sporangium discharge
- prepared slide of *Sphaerotheca panosa* conidia on rose leaf
- prepared slide of *Penicillium* conidia
- prepared slide of *Aspergillus* conidia
- cultures of a variety of cheese-flavoring fungi
- dissecting microscope
- compound microscope

PROCEDURE

1. Sporangia

   Zygomycetes reproduce asexually by forming sporangia. In this type of development numerous spores are made within a single cell, the spor-angium. A fungal sporangium is formed at the end of an aerial hyha called a **sporangiophore**. The sporangiophore tip swells to form a spor-angium, cell walls form around the numerous nuclei trapped within the sporangium, and, upon release, mitosis produces spores that germinate and grow into new mycelia.

   a. *Rhizopus stolonifer*

      To study asexual reproduction in a zygomy-cete, first examine the demonstration of the black bread mold, *Rhizopus stolonifer*. Note the dark specks covering the substrate surface.

The specks are sporangia filled with asexually produced spores. To supplement the living material, obtain a prepared slide of *Rhizopus* and use a compound microscope to study individual sporangia (Figure 16-7). Using low power plus scanning across the slide reveals sporangia at different stages of development. Locate a mature sporangium with spores. Also, note the coenocytic condition of the hyphae.

   b. *Pilobolus* spore dispersal

      Although most zygomycete asexual spores are wind dispersed, this demonstration shows an active means of zygomycete spore dispersal in the genus *Pilobolus*. Light directs the growth of the sporangiophores, and the sporangia are discharged as a mass. To see sporangiophores that have not yet fired their sporangia, take the lid off the culture dish and, using a dissecting microscope, look for surviving sporangia. Empty sporangiophores will be visible on the nutrient surface. Replace the lid after viewing, being careful to realign the lid as indicated.

2. Conidia

   Asexual reproduction in ascomycetes is by conidia formation. Conidia formation is very different from sporangia production. Conidia (or coni-diospores) are formed one at a time when a single cell, the conidium, is extruded from the tip of an upright hypha called a **conidiophore**. During conidia formation, the tips of hyphae produce one conidium at a time.

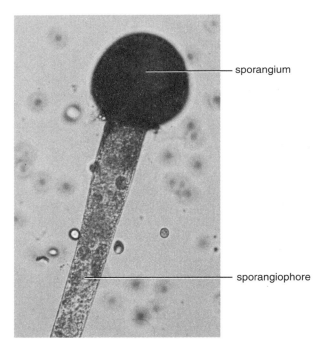

Figure 16-7 *Rhizopus* sporangium.

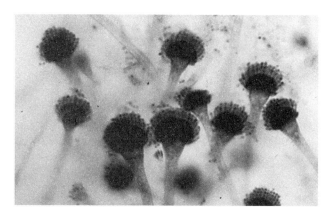

Figure 16-8 *Penicillium* conidia.

Examine the prepared slide of *Sphaerotheca panosa*, the organism that causes rose powdery mildew. Powdery mildews are obligate parasites that are host specific. They produce a mealy, powdery growth on leaf surfaces. Powdery mildews are particularly active during humid conditions and will continue to produce conidia over the growing season. Near the end of the growing season, the organism will reproduce sexually to make ascospores.

3. Division Deuteromycota—The Imperfect Fungi

This artificial division contains fungi that are known only to reproduce asexually. Most members form conidia; therefore, they are probably ascomycetes. Some grow in stored food and produce toxic substances, such as aflatoxins produced by *Aspergillus* species. Many of these organisms are beneficial and are used commercially—for example, *Aspergillus* species in large-scale production of citric acid and soy sauce, *Penicillium* in antibiotic manufacture, and *P. roquefortii* and *P. camembertii* in cheese-making. *Penicillium* and *Aspergillus* are typical conidia-forming deuteromycetes. Examine the demonstration slides of *Penicillium* (Figure 16-8) and *Aspergillus* and note the difference in their conidiophores. Also, examine the display of cheese-flavoring fungi.

PART C FEATURES OF FUNGAL HYPHAE

Fungi are heterotrophic organisms, both saprophytes and **parasites**, but because of their cell walls, they cannot engulf solid food particles. They obtain food by rapidly growing into a food source, where the fungal hyphae secrete enzymes that digest food particles, such as starches and proteins, into smaller molecules, such as sugars and amino acids. The smaller molecules are actively taken up by the hyphae. In the Zygomycota, the feeding hyphae lack cross walls. The hyphae within the divisions Ascomycota and Basidiomycota are septate, which means they have cross walls. In the following exercises, you will examine two features of fungal hyphae: septation and acid secretion.

MATERIALS

- fungus "A" culture
- fungus "B" culture
- fungal acid secretion display
- microscope slides
- coverslips
- dH$_2$O in dropper bottle
- dissecting needles
- compound microscope

PROCEDURE

1. Septation

Before beginning this exercise, read the next two paragraphs.

Place two drops of water on opposite ends of a clean microscope slide. Following the detailed instructions in the next paragraph, place a small amount of hyphae from the petri dish marked *"fungus A"* in the drop of water on the left side of the slide. The drop of water on the right side of the slide is for hyphae from *"fungus B."*

In order to see septation, or lack of septation, it is important to use young hyphae. Being careful not to pick up agar, remove a tiny portion of hyphae from the growing edge of each of the two mycelia in the petri dishes provided and place these in the separate drops of water. Use a probe to gently tease apart the hyphae in the drops of water (use a clean probe for each drop of water). Add coverslips and observe with the compound microscope.

Use low power to locate a region with hyphae and then change to high power to look for septation. First look at the hyphae in the drop of water on the left side of the microscope (fungus "A"), then observe the hyphae on the right side (fungus "B"). Can you find cross walls? Does fungus "A" have septate hyphae? Does fungus "B" have septa? How would you identify a "cell" in these organisms?

2. Acid secretion

Like fungi, many heterotrophic bacteria are decomposers. As a result, fungi and bacteria compete for some of the same food sources. However, as fungi grow they acidify the food source. Acidifying the growth medium improves the ability of fungi to successfully compete with bacteria that do not grow well at low pH.

Look at the demonstration of fungal acid secretion on the side bench. The fungus is growing on agar with an added pH indicator. The indicator is yellow when basic and turns pink when acidic. The sites of acid secretion are indicated by a color change from yellow to pink in the agar.

PART D FUNGI AND SYMBIOTIC ASSOCIATIONS

Fungi from all divisions are involved in numerous symbiotic relationships with algae, plants, and animals. We will discuss two fungi-plant associations in today's laboratory session. One is a mutualistic association between fungi and plant roots called a mycorrhiza. The other is a specific parasitic relationship called stem rust of wheat. We will examine a third type of association, lichens, in a later laboratory exercise.

Evidently the majority of vascular plants have mycorrhizal associations with fungi. These associations are beneficial to the plants because they increase mineral absorption. The two major types of mycorrhizae are ectomycorrhizae and endomycorrhizae. Basidiomycetes are the most common fungi involved ectomycorrhizae, and in these associations the fungus forms mantles of hyphae surrounding plant roots. In ectomycorrhizae, the fungus is located within the intercellular spaces of the root tissues. In endomycorrhizae, the most common type of mycorrhizal association, the fungal hyphae grow within plant cells. Most endomycorrhizal fungi are ascomycetes.

The rusts and smuts are basidiomycetes that cause diseases in plants. *Puccinia graminis*, commonly called stem wheat rust, is a nonbasidiocarp-producing basidiomycete and an important crop pathogen. Rusts illustrate a complicated life cycle that requires two hosts (barberry and wheat) to complete the sexual life cycle. A knowledge of a parasite's life cycle may be valuable in the human battle against pathogens.

MATERIALS

- prepared slide of *Corallorhiza striata* with endomycorrhiza
- prepared slide of *Puccinia graminis*: aecia
- prepared slide of *Puccinia graminis*: uredinia
- prepared slide of *Puccinia graminis*: telia
- compound microscope

PROCEDURE

1. Mycorrhizae

   Examine the prepared slide of *Corallorhiza striata* with an endomycorrhiza. Note the presence of fungal hyphae within the plant cells.

2. *Puccinia graminis*—a plant parasite

   Use a compound microscope to examine the prepared slides, looking for **aeciospores**, **uredospores**, and **teliospores** and noting which tissue the specific stage infects.

   a. Stages of development occurring on barberry

      Basidiospores are released in the spring. If basidiospores of compatible + and − mating types germinate on a barberry leaf, hyphae grow into the leaf tissue and eventually fuse to form dikaryotic hyphae that produce dikaryotic spores called aeciospores. Aeciospores are wind dispersed. Examine the slide with infected barberry leaf tissue and look for hyphae and clusters of aeciospores.

   b. Stages of development occurring on wheat

      If an aeciospore comes into contact with a wheat stem, it germinates and produces asexual, dikaryotic conidia called uredospores. During summer, uredospores continue to reinfect wheat. Clusters of uredospores on the surface of a wheat stem appear red, hence the name wheat stem rust. Near the end of summer, two-celled dikaryotic teliospores are produced in clusters that appear black on the wheat stem. Teliospores overwinter in the soil litter and complete the sexual life cycle by producing basidia with haploid basidiospores in the spring. Examine the wheat stem slides and locate uredospores and teliospores.

INTRODUCTION TO FUNGUSLIKE PROTISTS

Within the artificial kingdom Protista, there are divisions with photosynthetic organisms and divisions with heterotrophic organisms. The four heterotrophic divisions considered in this course are traditionally studied within the field of mycology. However, current evidence indicates that, except for the division Chytridiomycota, these taxa are not related to fungi. Chytrids are coenocytic, aquatic organisms and are thought the most likely ancestor of fungi. Like fungi, chytrids have cell walls of chitin; but, like motile cells in animals, chytrids are motile cells with a single, trailing flagellum. The division Oomycota contains unicellular to coenocytic filamentous organisms with cellulose cell walls. Some oomycetes are aquatic and known as water molds; others are terrestrial. The two divisions of slime molds are not related. One, division Aracsiomycota, contains the cellular slime molds; the other, division Myxomycota, contains the plasmodial slime molds. A cellular slime mold spends most of its life as a single, amoebalike cell with cellulose cell walls feeding on soil bacteria. If food supplies run short, cellular slime mold cells aggregate together, collectively forming reproductive structures and producing spores. We will briefly examine plasmodial slime molds and oomycetes.

PART A PLASMODIAL SLIME MOLDS

Plasmodial slime molds lack cell walls. The organism is a mass of coenocytic protoplasm that produces stalked sporangia when food is scarce. The feeding phase of a plasmodial slime mold can be extensive, and it is easy to observe internal cytoplasmic streaming.

MATERIALS

- culture of *Physarum*
- prepared slide of *Stemonitis* sporangia
- dissecting microscope
- compound microscope

PROCEDURE

1. Living *Physarum*

   Use a dissecting microscope to view the body of the plasmodial slime mold *Physarum*. Transfer the petri dish with *Physarum* to the stage of a compound microscope, remove the lid, and observe at low power. Note the rapid cytoplasmic streaming within the extensive protoplast.

2. *Stemonitis* sporangia

   Examine the prepared slide of sporangia of *Stemonitis*. Compare these structures with those seen in true fungi.

PART B OOMYCETES AND PLANT PATHOLOGY

Two terrestrial genera of oomycetes, *Plasmopara* and *Phytophthora*, cause much economic loss to plant disease. *Plasmopara* causes downy mildew on grapes, and *Phytophthora* species parasitize avocado, cacao, tomato, onion, apple, strawberry, and citrus crops. *Phytophthora infestans* is the agent of late blight of potato, a plant disease that ruined potato crops and resulted in massive starvation in 19th century Ireland.

MATERIALS

- prepared slide of *Phytophthora infestans* sporangiophores on leaf
- compound microscope

PROCEDURE

Using a compound microscope, examine the prepared slide of a potato leaf infested with *Phytophthora infestans*. The oomycete hyphae grow throughout the leaf, drawing energy to quickly reproduce by making sporangiophores and sporangia on the underside of the leaf. Note the extent of hyphae within the leaf tissue and the sporangiophores and sporangia.

QUESTIONS FOR THOUGHT AND REVIEW

1. What does the term aseptate mean? _____
   _____
   _____

2. Where does meiosis occur in zygomycete fungi?
   _____

3. Within an ascus, are ascospores produced by meiosis or mitosis? Explain. _____
   _____
   _____
   _____

4. How can asci be distinguished from hyphae within an ascocarp? _____

_____

5. Are the ascocarps produced by *Ascobolus* apothecia, perithecia, or cleistothecia? _____

6. List three commercial products that require yeast.

_____

_____

_____

7. What does n+n mean? Give two examples of n+n hyphae. _____

_____

_____

_____

8. List any evidence that sporangia of *Pilobolus* are shot with force. _____

_____

_____

_____

9. List several useful products derived from the fungi. _____

_____

_____

_____

10. Complete Table 16-1 summarizing some of the traits of fungi.

Table 16-1   Some Traits of Fungi

Division	Hyphae septation	Feeding hyphae n or n+n	Asexual reproduction	Sexual reproduction
Zygomycota				
Ascomycota				
Basidiomycota				

EXERCISE 16
LABORATORY QUIZ

Name:_____

Section Number: _____

FUNGI AND FUNGUSLIKE PROTISTS

1. Are fungi plants? Explain your answer._____

_____

2. What substance is found in both fungal cell walls and certain animals?_____

3. List the three divisions of fungi that reproduce sexually.

    a. _____

    b._____

    c. _____

4. Compare the formation of sporangiospores and conidiospores. _____

_____

_____

_____

5. How is a conidiospore like a basidiospore? _____

_____

How are they different? _____

_____

6. By what division process or processes are ascospores produced? Explain._____

_____

_____

_____

7. Describe an active means of spore dispersal in fungi._____

_____

_____

_____

8. Name the funguslike protist that caused massive starvation in Ireland during the 19th century. _____

_____

EXERCISE 17

ALGAE AND LICHENS

OBJECTIVES

1. From a photograph or a specimen, identify diatoms, dinoflagellates, and red, brown, and green algae.

2. List the specific and distinguishing characteristics of diatoms, dinoflagellates, and red, brown, and green algae.

3. Describe unicellular, colonial, filamentous, and thalloid algal growth forms.

4. Discuss the significance of the green algae and the origin of land plants.

5. Compare the sporic life cycles of an alga with an isomorphic alternation of generations and an alga with a heteromorphic alternation of generations.

6. Identify and describe a lichen.

TERMINOLOGY

accessory pigment	hypotheca
agar	kelp
alginic acid	phycobiliprotein
blade	pyrenoid
carrageenan	sieve cell
coralline alga	silica
epitheca	spine
flagellum	sporophyte
flotation bladder	stipe
gametophyte	thallus
holdfast	

INTRODUCTION TO THE ALGAE

Algae are photosynthetic, eukaryotic organisms that are primarily aquatic and are classified in the kingdom Protista. The fossils of algae include the oldest of plantlike organisms; in fact, algae gave rise to the land plants. Algal morphology is diverse. Forms range from unicellular to colonial to filamentous to multicellular, and their morphology reflects organisms adapted to life in surface and deep waters, from both marine and freshwater ecosystems, and in terrestrial environments. Criteria used to distinguish among the numerous algal divisions include cellular organization, cell wall compounds, energy-storing products, and photosynthetic pigments. Photosynthetic pigments give algae their characteristic colors and, in several divisions, their names.

PART A DIVISIONS OF ALGAE

1. Division Bacillariophyta—the Diatoms

Diatoms are unicellular or colonial algae present in large numbers in both marine and freshwater environments. They are major contributors to the base of the food chain. Diatoms have green chlorophylls a and c; however, because of an accessory photosynthetic pigment called fucoxanthin, diatom chloroplasts usually appear golden-brown. With the exception of a few species that have flagellated male gametes, all diatoms lack **flagella**. Each diatom cell is formed of two parts or valves: an upper **epitheca** and a lower **hypotheca**. The hypotheca and the epitheca fit together like a box with a lid.

Bilateral symmetry and radial symmetry are typical shapes of diatoms, and these symmetries reflect the two major groups of diatoms: pennate and centric. However, diatom symmetry is revealed only when the cells are seen in valve view (Figure 17-1a, page 130). When viewed from the side, the girdle view reveals the attachment line between epitheca and hypotheca (Figure 17-1b). Because they are made of **silica**, diatom cell walls are glasslike. The silica-rich cell walls preserve readily, and diatoms are well represented in the fossil record.

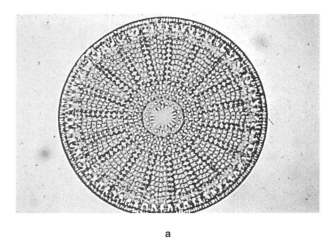

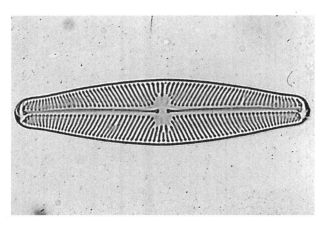

a

b

Figure 17-1 (**a**) Centric diatom. (**b**) Pennate diatom.

MATERIALS

- diatom type slide
- culture of *Pinnularia*
- diatomaceous earth
- microscope slides
- coverslips
- dH₂O in dropper bottle
- compound microscope

PROCEDURE

a. Diatom type slide

A diatom type slide is set up as a demonstration at a compound microscope. A diatom type slide displays rows of cleaned diatom shells exhibiting the diversity of cell shapes and wall patterns. You should be able to identify pennate and centric diatoms.

b. *Pinnularia* culture

Place a drop of the culture solution on a clean microscope slide, add a coverslip, and use a compound microscope to look for living *Pinnularia* cells. Try to find both valve and girdle views. Is *Pinnularia* pennate or centric?

c. Diatomaceous earth

Some fossilized diatoms are mined and used commercially. Scrape a small amount of diatomaceous earth into a drop of water on a microscope slide, cover, and observe under the compound microscope. Does the sample appear to contain many different species of diatoms? Are most pennate or centric? Inspect the display of products made with diatomaceous earth.

2. Division Pyrrhophyta—the Dinoflagellates

The majority of dinoflagellates (Figure 17-2) are photosynthetic unicellular algae with two flagella. Both flagella emerge from the same side of the cell, but one flagellum encircles the cell and the other trails backwards. Each cell has an upper and a lower half, and the outside is covered in a continuous cell membrane. Beneath the membrane are numerous polygonally shaped plates of flattened vesicles, called thecal plates, that fit tightly against each other in a pattern characteristic for each genus. In the so-called armored dinoflagellates, a structural polysaccharide such as cellulose fills the vesicles. Dinoflagellate chloroplasts normally appear brown because the yellow carotene and brown xanthophyll **accessory pigments** mask the green chlorophylls a and c. Dinoflagellates are an ancient group of algae and are important members of the phytoplankton. A few genera are known to cause the toxic red tides that result in fish kills and shellfish quarantines, and others are bioluminescent.

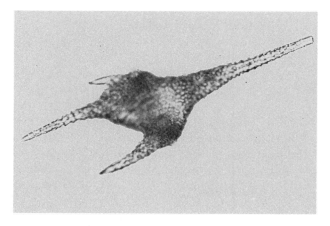

Figure 17-2 Dinoflagellate.

MATERIALS

- culture of *Peridinium*
- prepared slide of *Ceratium*
- microscope slides
- coverslips
- compound microscope

PROCEDURE

a. *Peridinium* culture

Make a wet mount of *Peridinium* from the living culture and use the high power objective lens of a compound microscope to look for the rapidly swimming cells. Describe their movement by looking for spinning and forward movement. Note the external pattern created by the thecal plates.

b. *Ceratium* prepared slide

Use a compound microscope to examine the prepared slide of the dinoflagellate *Ceratium*. Note the shape, and locate the horizontal and longitudinal grooves.

3. Division Rhodophyta—the Red Algae

Although the majority of red algae are marine organisms, a few live in freshwater. Red algae are particularly abundant in tropical marine waters and may grow at great depths in clear water. Their morphology ranges from unicellular to filamentous to bladelike. The most complex **thalli** of red algae consist of filaments compressed to form pseudoparenchymatous tissue. Many red algae are large enough to be called seaweeds. No red algal cells possess flagella, and chlorophyll a is the only chlorophyll present. Accessory pigments include the water-soluble **phycobiliproteins**: phycoerythrin, phycocyanin, and allophycocyanin. The cell walls of red algae may contain cellulose and the mucilaginous polysaccharides **agar** and **carrageenan**. The **coralline red algae** contain calcium carbonate in their walls, and some are important in reef formation.

MATERIALS

- prepared slide of *Batrachospermum*
- prepared slide of *Antithamnion*
- prepared slide of *Ceramium*
- living and herbarium specimens of marine red algae
- variety of commercial red algal products
- compound microscope

PROCEDURE

a. *Batrachospermum*

This genus is an example of a filamentous, freshwater red alga. It can be found in slow-moving streams and is known by the common name "frog-spawn" alga. It looks like soft, gelatinous beads on a string and is typically violet in color, although colors may range from brown to green. Use a compound microscope to examine the prepared slide and note the main filament with clusters of laterally branching filaments.

b. *Antithamnion*

Antithamnion is a filamentous, marine red alga. Examine the prepared slide with the compound microscope and note the simple, branching filaments. Compare this morphology with the more complex *Ceramium*.

c. *Ceramium*

Ceramium is an example of a more complex filamentous, marine red alga. In this genus, the thallus is made of numerous compressed filaments. Examine the prepared slide and compare it with the simpler *Antithamnion*.

d. Fresh and herbarium specimens of marine red algae

Examine the available living and herbarium specimens of marine red algae. Note any diversity in blade shape and color. Note how some of the living reds exhibit a blue and green iridescence (due to physical interference). If living specimens are available, feel the blades. Are they slimy or rough? Compare the texture of the coralline red algae with reds lacking calcium carbonate.

e. Commercial products from red algae

Agar and carrageenan are important commercial products extracted from red algae. Examine the display of products containing agar or carrageenan. Some red algae, such as *Porphyra*, are used for human food. You might know *Porphyra* by the name nori and have eaten it in sushi.

4. Division Phaeophyta—the Brown Algae

Brown algae are marine organisms ranging in size from microscopic, epiphytic filaments to the large seaweeds. The largest photosynthetic marine organisms, the **kelps**, are brown algae. Brown algae are common along rocky shores from zones continually submerged to zones exposed at low tide. The greatest diversity of these organisms occurs in the cool waters of temperate to subpolar regions. Their cell walls are made of cellulose and the mucilaginous polysaccharide **alginic acid**. The kelps have a complex **thallus** divided into **holdfast**, **stipe**, and **blade**. Within the stipe are specialized sugar-conducting cells called **sieve cells**. Many kelps have **flotation bladders**.

MATERIALS

- prepared slide of *Macrocystis* stipe
- living and herbarium specimens of marine brown algae
- variety of commercial brown algal products
- compound microscope

PROCEDURE

a. *Macrocystis* stipe

To see columns of sieve cells, examine the *Macrocystis* stipe slides. Look for sieve cells in the central region of the stipe, examining both cross and longitudinal sections. These cells are sometimes called trumpet hyphae. Look for cell traits that are "trumpetlike" or "hyphalike."

b. Living and herbarium specimens of marine brown algae

To appreciate properties contributing to the survival of brown algae on wave-pounded rocks, pick up the specimens and note the flexibility of the thallus. Describe the feel and texture of the surface of these marine organisms.

c. Commercial products from brown algae

Alginic acid is harvested from the cell walls of kelps and used commercially as a stablizer in products such as cosmetics and food products. Examine the display of products containing alginic acid.

---

5. Division Chlorophyta—the Green Algae

Although most green algae are freshwater organisms, many are marine or terrestrial. They can be found from the sea to lakes and streams to mountain snow fields. There are thousands of different species of chlorophytes, and their forms range from motile, unicellular organisms to macroscopic, thalloid forms. There are three major classes of green algae: the Chlorophyceae, the Ulvophyceae, and the Charophyceae. The classes are based on several complex features of biochemistry and cell structure. The motile cells of green algae typically have two flagella per cell. Like the red algae and the land plants, green algae chloroplasts are enclosed within one double membrane. (Algae in the other divisions have chloroplasts surrounded by one or two additional membranes.) Starch is the important reserve polysaccharide, and grains of starch are stored within the chloroplasts—a trait shared with land plants.

MATERIALS

- culture of *Pandorina*
- culture of *Volvox*
- culture of *Scenedesmus*
- culture of *Hydrodictyon*
- culture of *Ulothrix*
- living or harbarium specimen of *Ulva*
- culture of *Closterium*
- culture of *Micrasterias*
- culture of *Spirogyra*
- culture of *Chara*
- culture of *Coleochaete*
- I₂KI stain
- microscope slides
- coverslips
- compound microscope
- dissecting microscope

PROCEDURE

a. Chlorophyceae:

Pandorina

Pandorina is a colony of green algal cells in which each individual cell of the colony resembles *Chlamydomonas* (Figure 17-3). Take a drop of solution from the top of the culture tube. Deep in the solution, oxygen-deficiency causes cells to lose their flagella. Make a wet-mount slide and, starting with the low-power objective lens, locate a colony. Change to a high-power objective lens for a better view of individual colonies. Try to determine how many cells make up the colony and if that number appears to be constant.

Volvox

These colonies also consist of individual cells resembling *Chlamydomonas*. *Volvox* colonies are so large they will be squashed unless you examine them without a coverslip. An alternative is to place slivers of broken coverslip on the slide to raise the coverslip. Make a wet mount of *Volvox* and examine the preparation

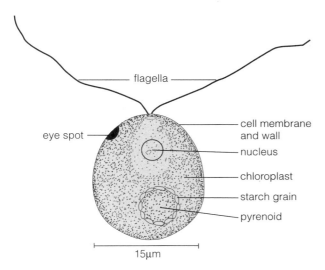

flagella

eye spot

cell membrane and wall

nucleus

chloroplast

starch grain

pyrenoid

15μm

Figure 17-3 *Chlamydomonas.*

at low magnification. Note juvenile colonies inside the large, parent *Volvox* colonies. Each juvenile colony started as the protoplast inside a single large cell, with all of the cells in the young colony forming before the colony breaks out of the parental cell wall. Many of the young colonies you see will soon break out of their parental cells, and then out of the parental colony. Note how *Volvox* moves about and look for a pattern of movement. If the colony appears to have a front end, look for differences in cell size from front to rear.

Scenedesmus

Scenedesmus is a nonmotile colony. It is one of many planktonic green algae that stay near water surfaces by catching upward currents with spinelike cell wall extensions. The presence of **spines** is a type of form resistance that increases drag and retards sinking rates. Make a wet mount, and examine at high power. Count the number of cells in a single colony and note the location of the spines.

Hydrodictyon

This is a beautiful green alga common in freshwater. The individual cells are large and cylindrical in shape. Cells are joined together in a polygonal configuration that produces a netlike thallus. Make a wet mount and first examine at low power to see the netlike form. Change to high power to examine the individual cells. Note how cells are interconnected.

b. Ulvophyceae:

Ulothrix

Ulothrix is a simple, unbranched filamentous green alga. The basal cell is modified as a holdfast that attaches the filament to a substrate. Make a wet mount and examine the preparation with a compound microscope.

Ulva

Inspect the demonstration of living or preserved *Ulva*, a green seaweed. *Ulva* is a typical thalloid marine chlorophyte and grows on rocks exposed at low tide. Note the sheetlike photosynthetic blade and the small holdfast. Compare the texture of *Ulva* with that of the kelps and the red and brown seaweeds.

c. Charophyceae:

Spirogyra

The large cylindrical cells of these unbranched filaments offer a good chance to see organelles. Adjust the microscope illumination carefully, and you may see a pale gray nucleus, with its spherical nucleolus, suspended by cytoplasmic strands in the center of the cell. The ribbon-shaped chloroplast (there may be more than one) is a green spiral that runs the length of the cell. To see **pyrenoids** within the chloroplast, apply I_2KI solution along the edge of the coverslip. Pull the I_2KI under the coverslip by wicking the solution with a bit of paper towel held at the opposite edge of the slip. The I_2KI solution will darken the pyrenoids, which are covered by starch grains and contain the enzyme ribulose bisphosphate carboxylase (RuBP carboxylase). RuBP carboxylase catalyzes the binding of CO_2 to RuBP at the start of the Calvin cycle. RuBP carboxylase is probably the most abundant protein on our planet.

Desmids

Desmids are strikingly beautiful unicellular green algae that are typically composed of two halves that are mirror images connected by a narrow isthmus. Make a wet mount and observe individual cells with the 40× objective. If you are viewing *Closterium*, which does not have an isthmus, look at the both ends of the cell for the lively movement of crystals within vacuoles. *Micrasterias* does have an isthmus, and it has characteristic deep grooves in the two cell halves.

Chara

Chara grows in mud or sand at the bottom of clear lakes and ponds, and it grows in streams. This green alga is well-branched, appearing to have nodes and internodes. *Chara* grows in a regular pattern because the filaments are produced from an apical cell. Examine living material with the dissecting microscope. Note the plantlike appearance of this alga.

Coleochaete

Green algae are thought to be the ancestors of land plants. Like plants, green algae have the photosynthetic pigments chlorophylls a and b and store starch in plastids. In addition, some members of the green algae (the family Charophyceae) have additional characteristics in common with plant cells, such as the position and anchoring characteristics of flagella in motile cells. During cytokinesis most organisms divide by furrowing, but plant cells form a cell plate at the cell equator. Some members of the Charophyceae, including the genus *Coleochaete*, divide by forming a cell plate. *Coleochaete* shares another important characteristic with plants: egg cells and zygotes are retained and sheltered on the parental body. In other algae, egg cells are released to drift away in water. The sheltering of the egg and zygote on the parental body is a very plantlike quality and leads us to consider *Coleochaete* an excellent model of a land plant ancestor. *Coleochaete* grows as small epiphytic discs or filaments, usually attached to other algae. To examine *Coleochaete* make a wet mount from the available culture. Note the shape of the thallus and look for setae. Setae are hairs extending from cells of the thallus and are characteristic of *Coleochaete*. Look for egg cells or zygotes.

PART B TWO TYPICAL LIFE CYCLES OF ALGAE

All three types of life cycles, zygotic, gametic, and sporic, are found in the algae. In a zygotic life cycle, the cells of the adult body are haploid. In a gametic life cycle, cells of the adult body are diploid. In organisms with a sporic life cycle, there is an alternation between two adult forms. One form has haploid cells and the other diploid cells. In this part of the exercise, you will compare two different types of sporic life cycle. In the first, the haploid and the diploid bodies are identical. This type of sporic life cycle is called an alternation of isomorphic generations. The second example is of an alga with an alternation of heteromorphic genera-

tions—the haploid and diploid bodies are different in appearance. Generally, in organisms with sporic life cycles the haploid generation is called a **gametophyte** and the diploid generation is called a **sporophyte**. Gametophytes produce gametes and sporophytes produce spores.

MATERIALS

- prepared slide of *Ectocarpus* with plurilocular sporangia
- prepared slide of *Laminaria* blade with sporangia
- compound microscope

PROCEDURE

1. Alternation of Isomorphic Generations

   Ectocarpus is a small, filamentous brown alga, and it will serve as an example of an alga with a sporic life cycle alternating between isomorphic haploid and diploid generations. The diploid thallus produces two different types of spores, one type by meiosis and one type by mitosis. The diploid thallus produces haploid spores by meiosis inside spherical cells called unilocular (or meiotic) sporangia. Upon release from a unilocular sporangium, the meiospores settle on a substrate and germinate to produce a new haploid generation. The diploid thallus produces diploid spores by mitosis inside an elongated plurilocular sporangium (or mitotic sporangium), and when these are released they grow into more diploid thalli. The haploid thallus also reproduces by mitosis inside elongated structures. However, on the haploid thallus, the cells released are gametes and the structure that produces them is called a gametangium. The gametes swim about and eventually fuse to form the new diploid cell, which germinates and grows into a new diploid thallus. Use a compound microscope to examine the prepared slide of *Ectocarpus* with plurilocular sporangia.

2. Alternation of Heteromorphic Generations

   Laminaria is a kelp and is a good example of an alga with an alternation of heteromorphic generations. All kelps have this type of life cycle. The haploid generation is very small and inconspicuous. The large organism we call the kelp is the diploid generation, and it is this generation that produces spores by meiosis. Obtain a prepared slide of *Laminaria* blade with sporangia and examine the sporangia with a compound microscope. Many sporangia are clustered together along the surface of the blade. Inside each sporangium, the diploid nucleus undergoes meiosis followed by mitotic divisions to produce as many as 64 haploid meiospores. The haploid meiospores settle to a substrate, germinate, and grow into the haploid generation that produces gametes. Upon fusion of the gametes, a zygote is produced. The zygote undergoes many mitotic divisions to mature into a new adult diploid generation.

INTRODUCTION TO LICHENS

Lichens are commonly encountered symbiotic associations between fungi and either cyanobacteria or green algae. Ninety-eight percent of the lichen-forming fungi are ascomycetes, and a few are basidiomycetes or deuteromycetes. The fungus gains photosynthates (products of photosynthesis) from the photosynthetic partner, and the photosynthetic alga or bacterium gains water and mineral nutrients from the fungal partner. Because lichens absorb nutrients and minerals from the air, they are particularly sensitive to air pollution.

MATERIALS

- display of a variety of lichens
- prepared slide of *Physcia* thallus
- compound microscope

PROCEDURE

A variety of lichens are displayed. Note their morphological differences. Lichens are classfied as crustose, foliose, or fruticose. Crustose lichens are very flat and grow close to the substrate. Foliose lichens also grow close to their substrate, but portions of the thallus fold away from the substrate. Fruticose lichens have erect growth, often with stalks of erect tissue that bear fungal reproductive structures. Look for any fungal reproductive structures. Examine the prepared slide of *Physicia* thallus with a compound miroscope. Identify fungal hyphae and algal cells within the fungal mycelium.

QUESTIONS FOR THOUGHT AND REVIEW

1. In which kingdom are algae classified?

2. What characteristic or combination of characteristics is unique to the diatoms?

3. What are the shapes of pennate and centric diatoms in top view? What are the shapes of these diatoms in side view?

4. What characteristic or combination of characteristics is unique to the dinoflagellates?

5. What trait of red algae allows them to photosynthesize at greater depths than other marine algae?

6. What is a kelp? In kelps, what are the functions of holdfasts, stipes, blades, and bladders?

7. Why are specialized sugar-conducting cells present in kelps, but specialized water-conducting cells not present?

8. What properties of intertidal algae make them well adapted to their habitat? Characterize any features or substances that provide these properties.

9. Given a container that lists an algal product among the contents, identify the material that comes from the alga and name the algal division from which the material was derived. _____

_____

10. Use Table 17-1 to summarize the listed traits of the red, brown, and green algae, and the diatoms and dinoflagellates.

Table 17-1 Some Agal Traits

Division	Storage product	Cell wall	Chlorophylls	Accessory pigments
Bacillariophyta				
Pyrrhophyta				
Rhodophyta				
Phaeophyta				
Chlorophyta				

EXERCISE 17
LABORATORY QUIZ

Name: _____

Section Number: _____

ALGAE AND LICHENS

1. What is the primary photosynthetic pigment in all algae?_____

2. List two accessory photosynthetic pigments in the red algae.

    a. _____

    b. _____

3. A typical kelp body is composed of what three parts?

    a. _____

    b. _____

    c. _____

4. Why do certain kelps have sieve cells but not vessel members?_____

    _____

    _____

5. List three reasons the Chlorophyta is considered the ancestral taxon to land plants.

    a. _____

    b. _____

    c. _____

6. What special substance is located in the cell walls of diatoms?_____

7. What is the function of alginic acid in brown algae?_____

    _____

    _____

8. List the two components of a lichen.

    a. _____

    b. _____

EXERCISE **18**

KINGDOM PLANTAE: BRYOPHYTES

OBJECTIVES

1. Compare the life cycle stages of liverworts, hornworts, and mosses.

2. Locate where meiosis occurs in bryophytes.

3. Compare the sterile jackets and reproductive cells of antheridia, archegonia, andsporangia in liverworts, hornworts, and mosses.

4. Describe means of spore dispersal in liverworts and mosses.

5. Identify what features present in modern bryophytes reveal their aquatic ancestry.

TERMINOLOGY

antheridium	hydroid
archegonium	leptoid
bulbil	operculum
calyptra	peristome
capsule	phyllid
caulid	protonema
elater	rhizoid
gametangium	sporangium
gametophore	sporophyte
gametophyte	sterile jacket
gemma	thallus
gemma cup	venter

INTRODUCTION TO THE SIMPLEST LAND PLANTS—THE BRYOPHYTES

Over 400 million years ago, the ancestors of modern plants left their aquatic environment and colonized the terrestrial environment. This monumental event raises an important question: which aquatic organisms were the ancestors of land plants? Because the photosynthetic, protistan algae are the most probable plant ancestors, one approach to solving this puzzle is to look for characteristics shared by modern plants and the different groups of algae. Comparing different traits within these groups reveals the green algae as the algal group with the most characteristics in common with plants and the most likley land plant ancestors. Specific characteristics used for such a comparison include types of accessory photosynthetic pigments, food storage compounds, cell wall materials, events of mitosis and cytokinesis, and traits of flagella in motile cells. For example, both green algae and land plants use chlorophyll b as an accessory photosynthetic pigment, and both groups of organisms store starch inside cell organelles called plastids. But compared with a watery habitat, the dry land is a harsh environment, and to survive in these harsh conditions land plants evolved new characteristics not present in green algae.

Among the new traits that allowed plants to adapt to the dry environment of terrestrial life are multicellular structures called **sterile jackets**. These structures surround and protect reproductive cells from desiccation, a major problem of life on land. Sterile jackets occur in two basic types, depending upon the kinds of reproductive cells they contain. If the reproductive cells are gametes, the multicellular structures and their gametes are called **gametangia**. There are two types of gametangia: **archegonia** contain egg cells and **antheridia** contain sperm cells. If the reproductive cells are spores, the multicellular structures containing spores are called **sporangia**.

Another adaptation to life on land is a waxy covering called a cuticle. Most plant bodies are covered by a cuticle, which reduces water loss from a plant's surface. But a covering that reduces water loss also decreases gas exchange required for photosynthesis. Along with a cuticle, plants evolved specialized epidermal structures called stomata that open and close to control CO_2 uptake and water loss.

Recall from Exercise 10 that plants have a sporic life cycle with two generations. Like all plants, bryophytes produce both **gametophyte** and **sporophyte** generations; however, unlike all other plant taxa, bryophyte gametophytes are the dominant generation of the life cycle. Bryophyte gametophytes are called dominant because they are conspicuous, photosynthetic, and long-lived; bryophyte sporophytes are not dominant because they remain attached to and nutritionally dependent upon their gametophytes. Being dominant does not mean that gametophytes are large, shrublike plants. In fact, lacking true vascular tissue (xylem and phloem), bryophyte gametophytes are limited in size. A plant body lacking xylem does not have true roots, stems, or leaves and is called a **thallus** (the plural is thalli). The bryophytes include plants commonly called liverworts, hornworts, and mosses.

PART A THE LIVERWORTS— CLASS HEPATICOPSIDA

Liverwort gametophytes grow from a single apical cell that produces a dorsiventrally flattened, dichotomously branched thallus. Gametophytes are either thallose or leafy. A thallose liverwort has a ribbonlike thallus; a leafy gametophyte has two to three rows of leaves. Since liverworts do not have vascular tissue, there are no true stems, leaves, or roots. Therefore, the "leaves" and "stems" of leafy bryophytes are sometimes called **phyllids** and **caulids**. Unicellular **rhizoids** grow from the lower surface of gametophytes, anchoring the organism to its substrate.

Liverworts do not have a true epidermis with stomata, instead open pores are present in the upper surface of the thallus. Within this group the position of reproductive structures is variable.

MATERIALS

- variety of living thallose and leafy liverwort gametophytes
- living *Marchantia* gametophytes with gemmae cups
- prepared slide of *Plagiochila* thallus
- prepared slide of *Marchantia* thallus
- living *Marchantia* gametophytes with antheridiophores
- living *Marchantia* gametophytes with archegoniophores
- living *Marchantia* gametophytes with mature sporophytes
- prepared slide of *Marchantia* antheridia
- prepared slide of *Marchantia* archegonia
- dissecting microscope
- compound microscope

PROCEDURE

1. Liverwort Habit

    Different liverwort species are displayed. Note the size and shape of each organism and look for similarities and differences in their morphology. To locate the growing tips of a thallus, examine a liv-

erwort with a dissecting microscope. To see a type of asexual reproduction in liverworts, look at the specimen of *Marchantia* noted as having **gemmae cups**. Gemmae cups contain small disks of thallus tissue called **gemmae**. When it rains, the gemmae are splashed out of the cups and grow into clones of the parent thallus. To get a better look at gemmae, examine the thallus under a dissecting microscope.

2. Thallus Anatomy

    a. A leafy liverwort

    Examine the cross section of the leafy liverwort *Plagiochila*. Start with low power of a compound microscope and then move to high power. Note the simplicity of the tissue and the lack of vascular tissue. Compare this liverwort thallus with the thallose liverwort cross section.

    b. A thallose liverwort

    Use a compound microscope to examine the cross section of a *Marchantia* gametophyte thallus (Figure 18-1). Thalloid liverworts have a more complex anatomy than leafy liverworts. However, note that the thallus lacks vascular tissue. As you look at the tissue visible in the cross section, speculate about the functions of cells with different shapes and sizes, locate pores in the upper thallus surface, and identify rhizoids at the lower surface of the thallus.

3. Sexual Reproduction

    a. Antheridia

    Using a compound microscope, examine a *Marchantia* prepared slide with antheridia. In *Marchantia*, antheridia differentiate in special-

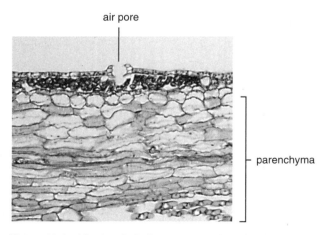

Figure 18-1 *Marchantia* thallus, cross section, showing a pore in upper surface and simple parenchyma tissue of the thallus.

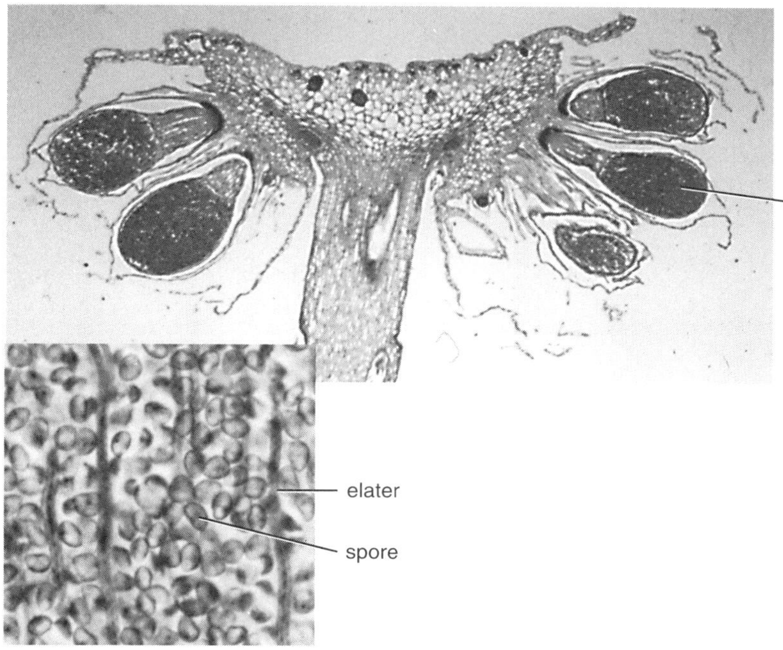

Figure 18-2 *Marchantia* archegoniophore, longitudinal section, with sporophytes at different stages of development.

spores (*n*) and elaters (2*n*)

elater

spore

ized structures called antheridiophores. Antheridiophores look like tiny umbrellas, and several antheridia are embedded in the top of each antheridiophore.

b. Archegonia

Use a compound microscope to examine *Marchantia* archegoniophores, which look like small palm trees, and locate the archegonia (Figure 18-2). Archegonia develop on the underside of the top portion of the archegoniophores.

4. Sporophytes

When mature, the sporophytes of *Marchantia* liverworts consist of three parts: a foot, a seta, and a **capsule**. The cells of the foot region remain embedded in the gametophyte tissue; the seta elongates into a slender stalk and is topped by the capsule. The capsule is the sporangium. When mature, the capsule contains two types of cells: spores and **elaters**. Within the mass of spores the elaters twist in response to changes in humidity and aid in spore dispersal. Mature *Marchantia* sporophytes are easy to identify; they contain bright yellow spores. With a compound microscope, examine the prepared slide of *Marchantia* sporophytes and locate the foot, seta, and capsule of a mature sporophyte. At high power, examine the contents of a capsule. You should be able to locate spores and elaters.

PART B **THE HORNWORTS—CLASS ANTHOCEROTOPSIDA**

Hornwort gametophytes resemble small thalloid liverworts with wavy edges, but their sporophytes are unique. The hornwort sporophyte superficially resembles a small, thin blade of grass. This elongated structure has a foot embedded in the gametophyte thallus and an indeterminate capsule—the capsule has an intercalary meristem at its base. New meiospores are continuously produced during favorable growing conditions.

MATERIALS

- living hornwort gametophyte
- prepared slide of *Anthoceros* gametophyte with attached sporophyte
- dissecting microscope
- compound microscope

PROCEDURE

1. Hornwort Habit

Examine the living hornwort gametophyte and compare it with thalloid liverworts. Look for developing sporophytes. Use a dissecting microscope for the best view of these small structures.

2. Sporophytes

Use a compound microscope to examine the prepared slide of an *Anthoceros* sporophyte. Locate a capsule. Each capsule has two halves divided by a column of tissue and contains maturing spores and elaters. Mature spores have dark, thick cell walls. As spores mature, the capsule dries and splits.

PART C THE MOSSES—CLASSES ANDREAOPSIDA, BRYOPSIDA, SPHAGNOPSIDA

Mosses are important as pioneer plants in ecological primary and secondary succession, and they are important as primary producers in a variety of terrestrial enivronments. Mosses are divided into three groups, frequently elevated to the class level. The Andreaopsida are the granite mosses, the Bryopsida are known as the true mosses, and the Sphagnopsida are the sphagnum or peat mosses. Except for the peat mosses, mosses have little economic value.

Mosses have upright, leafy gametophytes; multicellular, filamentous rhizoids; and leaves that are spirally arranged around the stem. Leaves usually have a distinct midrib. However, because they do not develop vascular tissue, the upright, leafy structures do not have true leaves. Some mosses have a central strand of specialized conducting cells called **hydroids** and **leptoids**. Moss water-conducting cells are thin-walled and dead at maturity, but unlike xylem conducting cells, the hydroids lack lignified cell walls. Outside the hydroids a cylinder of sugar-conducting cells, leptoids, may be present. Leptoids are similar to sieve cells. Unlike liverworts and hornworts, moss sporophyte morphology is highly diverse.

MATERIALS

- living moss gametophytes
- prepared slide of *Polytrichum* leaf, whole mount
- prepared slide of *Mnium* leaf, whole mount
- prepared slide of *Sphagnum* leaf, whole mount
- prepared slide of *Mnium* leaf, cross section
- prepared slide of *Sphagnum* leaf, cross section
- prepared slide of *Mnium* antheridia
- prepared slide of *Mnium* archegonia
- prepared slide of *Polytrichum* sporophyte
- living moss gametophytes with mature sporophytes demonstrating peristome action
- prepared slide of moss protonema with bulbils
- prepared slide of *Polytrichum* seta cross section
- dissecting microscope
- compound microscope

PROCEDURE

1. Moss Habit

Use a dissecting microscope to examine living moss gametophytes. Compare the living mosses on display, pay particular attention to the upright axes, leaf-like structures along the axes, and the rhizoids at the base of each axis. Compare the moss habit with the liverwort and hornwort habit. Look for developing sporophytes.

2. Leaf Morphology and Anatomy

Use designated living material and the prepared slides of whole mount leaves to compare a series of leaves from genera in the classes Bryopsida and Sphagnopsida. Use a dissecting microscope to see these details. Note the overall leaf shape, the leaf margin, and the position and extent of any midrib. Now, with a compound microscope, compare the anatomy of the *Mnium* and the *Sphagnum* leaves. Look for any differences in the number of cell layers, the shapes of the cells, and the thickness of the cell walls. All of these characteristics contribute to species determination.

3. Stem Anatomy—Hydroids and Leptoids

Obtain a slide of *Polytrichum* stem from the side bench. The slide will reveal three regions: an outer epidermal-like layer, a wide cortical region, and a central strand of cells. The central region is composed of thick-walled and thin-walled cells. The thin-walled cells in the center are hydroids, and they are surrounded by thin-walled leptoids. Hydroids conduct water and leptoids conduct photosynthate. Note the simplicity of this tissue when compared with the vascular stem tissue of flowering plants.

4. Sexual Reproduction

a. Antheridia

Using a compound microscope examine a prepared slide of *Mnium* antheridia. Antheridia are located at the top of each gametophyte (Figure 18-3). Each antheridium is composed of a multicellular sterile jacket enclosing and protecting the developing male gametes, sperm. Look at the gametes within the antheridia and try to determine their stage of development. When mature, antheridia release swimming sperm cells, which require free water to reach the egg-containing archegonia.

b. Archegonia

On a prepared slide labeled *Mnium* archegonia, locate the egg-producing gametangia (Figure 18-4). The archegonia are flask-shaped

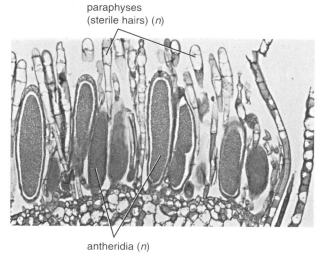

paraphyses
(sterile hairs) (*n*)

antheridia (*n*)

Figure 18-3 *Mnium* gametophyte, longitudinal section, with multiple antheridia.

and contain a single fertile cell—the egg cell. The egg cell is surrounded by the smaller, sterile cells of the archegonium; the enlarged, basal portion of the archegonium is called a **venter**. Archegonial neck cells extend from the venter, providing a path down which swimming sperm cells reach the egg cell.

The fusion of sperm and egg cell nuclei (syngamy) results in a zygote. The zygote, while retained within the archegonium, divides repeatedly, producing a spherical mass of cells, the embryo sporophyte. The mature sporophyte is composed of three regions: a foot, a seta, and a sporangium (or capsule).

paraphyses
(sterile hairs) (*n*)

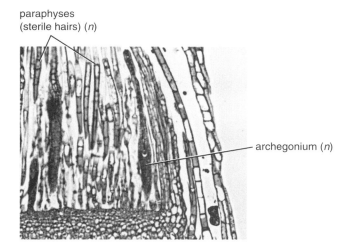

archegonium (*n*)

Figure 18-4 *Mnium* gametophyte, longitudinal section. An archegonium with an egg cell is visible.

5. Sporophytes

Examine any living specimens available. You may see a shaggy covering draped over the sporangium of the sporophyte. This covering is a **calyptra**, which developed when sterile cells in the old archegonium continued dividing and elongating after egg cell fertilization. Use a compound microscope to examine a prepared slide of *Polytrichum* capsules (Figure 18-5). Each immature sporangium consists of a sterile jacket of cells surrounding a mass of spore-producing tissue. During maturation, cells within the sporangium divide by meiosis to produce spores.

6. Spore Dispersal

Moss capsules actively disperse spores. When a moss capsule is mature, a cap of cells, the operculum, falls away to expose rows of teethlike cells, the **peristome**. Look at the peristome demonstration and note how the peristome teeth open and close with changes in humidity. When the atmosphere is dry, the peristome opens, lifting spores. The spores are then dispersed by wind.

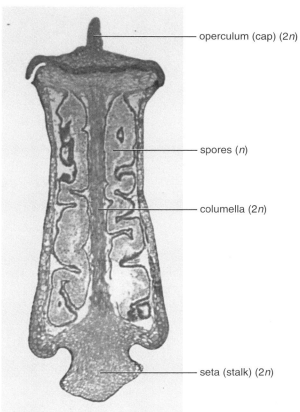

operculum (cap) (2*n*)

spores (*n*)

columella (2*n*)

seta (stalk) (2*n*)

Figure 18-5 Moss sporophyte, longitudinal section. Sporangium and a portion of seta are visible.

7. Spore Germination and Early Growth of the Gametophyte

Once spores land on a suitable substrate, they germinate. In the class Bryopsida a germinating spore immediately develops into a filamentous structure called a **protonema**. In the early stages of development, a protonema look like a filamentous green alga. Eventually, the filamentous protonema branches, sends down rhizoids, and develops numerous small buds called **bulbils**. The bulbils enlarge and grow upright, becoming the erect moss axis of stem and leaves. To distinguish the protonema stage from the upright stage of a gametophyte, the erect axis that produces antheridia and archegonia is sometimes called a **gametophore**.

QUESTIONS FOR THOUGHT AND REVIEW

1. What is the dominant generation of liverworts, hornworts, and mosses? _____

_____

_____

2. What is a rhizoid, and what is its function? Why are rhizoids not considered true roots? _____

_____

_____

_____

3. Are there visible pores on the surface of liverworts? Are these pores stomata? Why or why not?

_____

_____

_____

_____

4. Are there any obvious regions of growth in a liverwort thallus? Do the plants branch? If yes, speculate about the origin of the "branches" in the gametophyte body. _____

_____

_____

_____

5. Are the oldest spores in a hornwort capsule near the top or the bottom of the capsule? _____

_____

6. What shape are antheridia? What cells are produced within antheridia, and are they produced by mitosis or meiosis? _____

_____

7. What shape are archegonia? What cells are produced within archegonia, and are they produced by mitosis or meiosis? _____

_____

8. How might sterile jackets have been an adaptation to life on land? _____

_____

_____

_____

9. What is a capsule? How are liverwort and moss capsules specialized for spore dispersal? Does the peristome of a moss open or close with decreasing humidity? _____

_____

_____

_____

10. In mosses, what is the function of the calyptra?

_____

_____

_____

_____

EXERCISE 18
LABORATORY QUIZ

Name: _____

Section Number: _____

KINGDOM PLANTAE: BRYOPHYTES

1. Using their informal names, list the three classes of bryophytes.

    a. _____

    b. _____

    c. _____

2. Like all plants, bryophytes have a sporic life cycle. What is another, more descriptive name for this type of life cycle? _____

3. What is a gametophyte? _____

_____

4. Where are spores produced in a liverwort? _____

5. What is the purpose of elaters? _____

_____

_____

6. Bryophytes are nonvascular plants. What does this imply about their maximum potential size on land?

_____

_____

7. What is the function of a peristome? _____

_____

_____

8. In plants, what division process (mitosis or meiosis) produces egg cells and sperm cells? _____

_____

_____

EXERCISE 19

KINGDOM PLANTAE:
SEEDLESS VASCULAR PLANTS

OBJECTIVES

1. Recognize sporophyte stages of seedless vascular plants.

2. Recognize gametophyte stages of *Selaginella* and ferns.

3. Identify true leaves and roots.

4. State whether a given specimen belongs to the Psilophyta, Lycophyta, Sphenophyta, or Pterophyta.

5. Distinguish between homosporous and heterosporous sporangia and spores.

6. Understand the life cycle stages of a typical homosporous fern.

7. Understand the life cycle of the heterosporous seedless vascular plant, *Selaginalla*.

TERMINOLOGY

annulus	microphyll
heterospory	microsporangium
homospory	microspore
indusium	microsporophyll
megagametophyte	rhizome
megaphyll	sorus
megasporangium	sporangium
megaspore	sporophyll
megasporophyll	strobilus
microgametophyte	

INTRODUCTION TO THE SEEDLESS VASCULAR PLANTS

Although most seedless vascular plants (sometimes called the lower vascular plants) are inconspicuous today, during the Carboniferous period, which extended from about 360 to 286 million years ago, three divisions of seedless vascular plants—the lycophytes, the sphenophytes, and the pterophytes—formed the dominant vegetation on Earth. Land formations were low during the Carboniferous period,

the earth was covered with shallow seas and swamps, and the climate was mild. These conditions favored the formation of coal. Because of the large coal deposits laid down during the Carboniferous period, it is also known as the Coal Age. Because seed plants evolved from seedless vascular plants, living seedless vascular plants are useful in attempting to understand the origin of seed-producing plants such as pine trees and flowering plants.

A prerequisite to the evolution of the seed was the origin of two different sizes of spores. The liverworts and mosses examined in the previous exercise develop spores of a single size, a condition called **homospory**. Homospory is present in some of the seedless vascular plants, but others have two different spore sizes. Having spores of two different sizes is called **heterospory**. Heterospory arose in ancient seedless vascular plants, and, although seed plants did not evolve from the lycophyte group of seedless vascular plants, a living lycophyte, *Selaginella*, is easy to obtain and helpful as an introduction to heterospory.

The bryophyte exercise introduced the concept of an alternation of generations, something that all plants share. The haploid gametophyte generation produces gametes (egg and sperm cells) and the diploid sporophyte generation produces spores. Unlike bryophytes, which have dominant gametophytes, all plants with vascular tissue—xylem and phloem—have dominant sporophytes, and seedless vascular plants are the simplest plants with a dominant sporophyte generation. There are four divisions with living members, with ferns comprising the group with the greatest number of species.

PART A DIVISION PSILOPHYTA— THE WHISKFERNS

Although it is a vascular plant, the *Psilotum* sporophyte does not have true roots or leaves. The sporophyte body consists of an aerial stem system and a horizontal rhizome system, and both systems have vascular tissue. Rhizoids emerge from the underground rhizome system. Along the aerial stem are small epidermal outgrowths, sometimes called enations, that superficially look like scale leaves. However, enations lack vascular tissue, so they are not true leaves. Sporangia occur along the stems near enations. Because *Psilotum* is homosporous, its gametophytes are independent; however, they are inconspicuous (*Psilotum* gametophytes look like small segments of rhizome tissue) and grow underground in association with a mycorrhizal fungus. *Psilotum* is not economically important, and it lacks a fossil record.

MATERIALS

- living sporophytes of *Psilotum* with mature sporangia
- prepared slide of *Psilotum* stem cross section
- compound microscope

PROCEDURE

1. *Psilotum* Sporophyte Habit

    Examine both the aerial stem and the underground rhizome systems of a living *Psilotum* sporophyte (Figure 19-1). The leafless aerial stems are green and photosynthetic. The dichotomous branching pattern reveals a primitive architecture; dichotomous branching is common in seedless vascular plants. Note the shape and texture of the stems and the overall size of the plant. Also look at the underground rhizome system with rhizoids.

2. *Psilotum* Sporangia

    Scan the green stems for bright yellow sporangia, which are in tight clusters of three. A cluster of sporangia fused in this manner is called a synangium. The sporangia produce homosporous

spores, and at maturity the spores are released and germinate into gametophytes.

3. *Psilotum* Stem Anatomy

    Use a compound microscope to view the prepared slide of the *Psilotum* stem in cross section. Note the epidermis with stomata, cortex, and central region of vascular tissue. This central arrangement of vascular tissue is similar to that of flowering plant roots. If the section includes an enation, note that the enation lacks vascular tissue.

PART B DIVISION LYCOPHYTA— THE CLUBMOSSES

Living lycophytes are small plants and none has significant economic value, but the lycophytes of the Carboniferous period included tree-size forms with limited secondary growth, and their remains contributed to the formation of coal. All lycophytes have true stems, leaves, and roots. The leaves are small, and the stems branch more or less dichotomously. The division contains homosporous members and heterosporous members. Sporangia develop singly on the upper side of leaves near the stem axis and are usually clustered in terminal **strobili**, or cones. The gametophytes of homosporous lycophytes are similar in morphology to those of *Psilotum*; the gametophytes of heterosporous lycophytes are endosporic. (Gametophytes of homosporous plants are exosporic.) Endosporic means that the gametophyte is retained and matures within the spore wall. Thus, these endosporic gametophytes are always very small and never photosynthetic. The homosporous *Lycopodium* and the heterosporous *Selaginella* represent the division Lycophyta.

MATERIALS

- living sporophytes of *Lycopodium*
- prepared slide of *Lycopodium* strobilus in longitudinal section
- fossil specimen of *Lepidodendron*
- living sporophytes of *Selaginella*
- living *Selaginella* strobili producing microsporangia and megasporangia
- prepared slide of *Selaginella* strobilus in longitudinal section
- dissecting needles
- dissecting microscope
- compound microscope

Figure 19-1 *Psilotum* sporophyte. Habit view showing upright, aerial stems. Note dichotomous branching and sporangia.

Figure 19-2 *Lycopodium* sporophyte. Habit view showing stem, leaves, and sporangia.

PROCEDURE

1. *Lycopodium*

   a. Sporophyte habit

      Lycopodium is a homosporous lycophyte with true leaves and roots (Figure 19-2). Examine the stems and note the number of veins within each leaf. Generally, leaves with a single vein are called **microphylls**. Note the arrangement of leaves around the axis. Look for *Lycopodium* sporangia, which are located in the axils of specialized microphylls called **sporophylls**.

   b. Strobilus

      With a compound microscope, examine the prepared slide of a *Lycopodium* strobilus. The strobilus was cut longitudinally to reveal leaves bearing sporangia. The leaves are called sporophylls because they bear sporangia. Each sporangium contains a single size of spore; therefore, *Lycopodium* is homosporous.

2. *Lepidodendron* Fossil

   One of the more important seedless vascular plants in Carboniferous period coal beds is the lycophyte tree *Lepidodendron*. Its slender trunk and crown of dichotomously forking branches reached over 40 m in height. Lance-shaped leaves spiralled around the branches, and the abscised leaves left a characteristic pattern of leaf scars on the branches. The root system was shallow, so the trees may have been toppled easily by winds. Examine the fossils of *Lepidodendron* and note the distinctive leaf scar pattern.

3. *Selaginella*

   a. Sporophyte habit

      Inspect the living examples of *Selaginella* and take a piece of stem to your work space to examine with the aid of a dissecting microscope (Figure 19-3). Look for a pattern in the arrangement of leaves along the stem and look for differences in leaf size. Note the rhizome system with its emerging roots, and note the pale rootlike structures extending from stems. These are rhizophores. Rhizophores emerge near sites of branching and upon contact with the soil their tips develop roots.

   b. Strobili

   Living plants

   Use the same plants or plants set aside for this part of the exercise. If you have not already done so, get out a dissecting microscope; it will be impossible to see the following structures without it. Examine the ends of stems and look for strobili. *Selaginella* strobili are not always conspicuous. Pinch a strobilus from its stem and place it on the stage of a dissecting microscope. Use dissecting needles to carefully peel back a sporophyll, then look at the upper sporophyll surface. A sporangium should be exposed. *Selaginella* plants produce two kinds of sporangia, **microsporangia** and **megasporangia** (Figure 19-4, page 150).

Figure 19-3 *Selaginella* sporophyte. View of stem and leaf arrangement.

Kingdom Plantae: Seedless Vascular Plants **149**

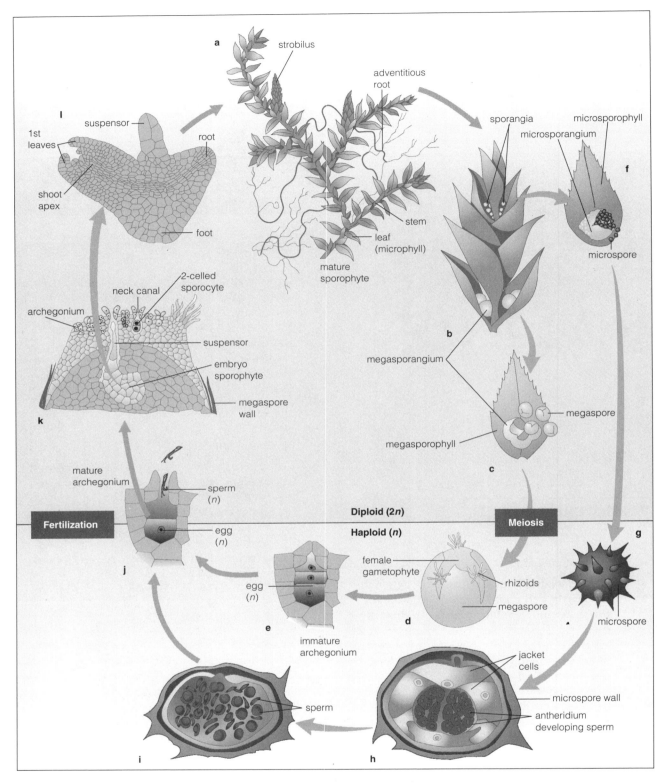

Figure 19-4 *Selaginella* life cycle.

Strobili may contain mixtures of microsporangia and megasporangia, but each sporophyll has only one type of sporangium. A sporophyll with **microsporangia** is called a **microsporophyll**; **megasporophylls** bear **megasporangia**. Mature microsporangia, which appear bright orange, produce many tiny **microspores**; and within each microspore a **microgametophyte** develops. Mature megasporangia are pale yellow and appear lumpy because they contain four large **megaspores**. Megaspores develop into **megagametophytes**.

Prepared slide

For a different view of *Selaginella* strobili, obtain a prepared slide with longitudinal sections. Using a compound microscope, identify micro- and megasporophylls, micro- and megasporangia, and micro- and megaspores. Closely examine a megasporangium and note the number of megaspores it contains. Look at a single megaspore and identify the spore wall.

c. Megagametophyte

Observe the demonstration of living *Selaginella* megagametophytes. Megagametophytes mature after they are released from megasporangia. Although megagametophytes remain enclosed wthin megaspore walls, the walls rupture to expose archegonia. To locate archegonia, look for clusters of neck cells protruding from the megagametophyte tissue surface. Note the small rhizoids extending from the megagametophyte tissue near the region of archegonia. Microgametophytes must be near by, and a film of water is required for the sperm to swim from the microgametophytes to the archegonia of the megagametophytes, to accomplish fertilization, and to produce a zygote.

PART C DIVISION SPHENOPHYTA— THE HORSETAILS

Equisetum architecture is different from the lycophytes, and these plants, although lacking secondary growth, can reach greater heights than living lycophytes. There are true leaves and roots, but the leaves are tiny and arranged in whorls around each node (Figure 19-5). Stems are the primary synthetic organs. If lateral branches are present, they also are arranged in whorls at the nodes. Branches and leaves alternate around each node. *Equisetum* is homosporous and sporangia are borne in distinctive, terminal strobili. Game-tophytes are photosynthetic. This is an ancient group of plants, and during the Carboniferous period the giant *Equisetum*-like plant *Calamites* was an important forest tree.

MATERIALS

- living sporophytes of *Equisetum* with strobili
- prepared slide of *Equisetum* strobilus, cross section and longitudinal section
- living *Equisetum* gametophytes
- prepared slide of *Equisetum* gametophyte with antheridia and archegonia
- fossil specimen of *Calamites*
- dissecting microscope
- compound microscope

PROCEDURE

1. *Equisetum*

    a. Sporophyte habit

    Examine the specimens available. Feel the rough, ribbed stems and note the whorls of alternating branches and leaves at the nodes. The stems of *Equisetum* are called jointed because of the alternating node-internode construction and the ease with which the internodes can be popped apart. If enough

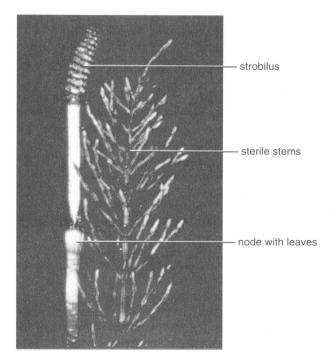

Figure 19-5 *Equisetum* sporophytes. Habit view showing jointed stem and strobili.

material is available, pull apart a stem. Look inside an internode to see the hollow stem.

b. Strobili

Note the complex structure of these strobili; clusters of sporangia extend from the strobilus axis. Because *Equisetum* is homosporous, the spores are all the same size. If a fresh strobilus is available, open it, remove a few spores and place them on a microscope slide, and examine them under high power of a compound microscope. The spores are photosynthetic and have two cell wall extensions, called elaters, which aid in spore dispersal by wind. When drifting over an area with relatively high atmospheric humidity, the elaters wrap around the spore, and it drops to the surface, ready to germinate in a moist region. Supplement the living material with the prepared slide of an *Equisetum* strobilus. Use a compound microscope to examine both cross and longitudinal sections.

c. Gametophyte

Equisetum gametophytes are independent and photosynthetic. If available, examine the living gametophytes with a dissecting microscope. Note the small size and the shape of the gametophyte thallus. Use a compound microscope to examine the prepared slide of an *Equisetum* gametophyte with antheridia and archegonia. The structure of antheridia and archegonia is similar to bryopyte and other seedless vascular plant gametangia.

2. *Calamites* Fossil

Look at the *Calamites* fossil on display. Like the living *Equisetum*, *Calamites* had an underground rhizome system and erect branches with whorls of alternating branches and leaves at the nodes. The fossil stem reveals the *Equisetum*-like ribbed surface and nodes of this Carboniferous period plant that reached heights of 18 m.

PART D DIVISION PTEROPHYTA— THE FERNS

As the largest and most diverse group of seedless vascular plants, ferns are important members in localized environments, and they have some economic value in the nursery trade. Most ferns have large, compound leaves, called **megaphylls**, and an extensive **rhizome** system. Most ferns are homosporous, but a few, the water ferns, are heterosporous. This exercise is restricted to the more familiar homosporous ferns. While examining the ferns on display, refer to the fern life cycle diagram (Figure 19-6).

MATERIALS

- living fern sporophytes with different soral types
- prepared slide of *Cyrtomium* leaf in cross section with sporangia
- prepared slide of *Adiantum* leaf in cross section with sporangia
- prepared slide of *Dennstaedtia* leaf in cross section with sporangia
- prepared slide of *Polypodium* leaf in cross section with sporangia
- living fern gametophytes
- prepared slide of fern gametophyte with antheridia and archegonia
- prepared slide of fern gametophyte with young sporophyte
- dissecting microscope
- compound microscope

PROCEDURE

1. Fern Sporophyte Diversity

Inspect and compare the different ferns available. Compare each living fern with the diagram in Figure 19-7, page 154. Locate the rhizome system of each specimen and note the complexity of its leaves. (An entire fern leaf is sometimes called a frond.) Identify the petiole, rachis, and leaflets. If present, note that young, unexpanded leaves are curled into "fiddlenecks." Fern leaves have many veins, and sporangia develop on the lower surface.

2. Soral Diversity

The majority of ferns develop **sporangia** in clusters on the underside of leaf tissue. A cluster of fern sporangia is called a **sorus**, and the shape and location is species specific. A selection of fern leaves showing different arrangements and shapes of sori is on display. Note the location of each sorus relative to the leaf margin and veins. Also, note whether individual sori are round or linear in shape. Many sori are covered with an umbrella-like structure called an **indusium**. The indusium extends over and protects immature sporangia, becoming dry and papery as the sporangia mature and release their spores. Indusium shape is variable.

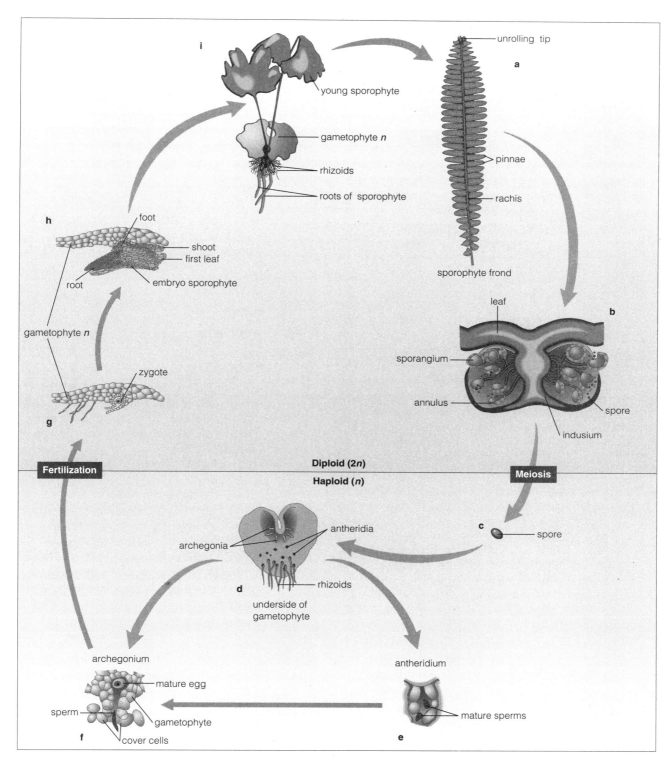

Figure 19-6 Fern life cycle.

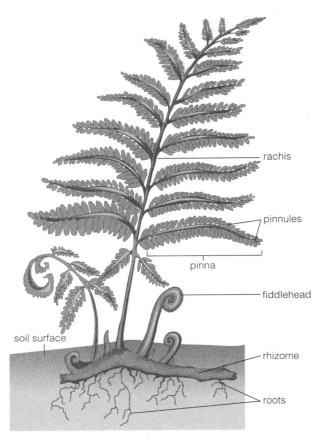

rachis

pinnules

pinna

fiddlehead

soil surface

rhizome

roots

Figure 19-7 Fern sporophyte.

3. Sorus Anatomy

a. *Cyrtomium*

Use a compound microscope to examaine the prepared slide of a *Cyrtomium* fern leaf cross section. First use a low power objective lens to scan across the leaf tissue and locate an intact sorus. Note the indusium and the numerous sporangia. Compare the appearance of the indusium in the prepared slide with the appearance of peltate-shaped indusia in living specimens. Change to a higher power objective lens and examine individual sporangia. Compare the various developmental stages of individual sporangia within a single sorus. Mature spores should be visible within some of the sporangia. Also note the specialized sterile jackets of fern sporangia. There are differences among the cells of the sterile

jacket. Some cells are thick walled, and the region is called the **annulus**. It aids in spore dispersal. When mature, the jacket wall of the sporangium splits and actively flings spores into the air. Place pieces of fresh leaf tissue with mature sporangia on the stage of a compound microscope under a bright, warm light. You may be rewarded with the sight of sporangia bursting open and flinging spores.

b. *Adiantium, Dennstaedia, Polypodium*

If available, compare the anatomy of these genera with *Cyrtomium*. Instead of a true indusium, the leaf margin of *Adiantium* curves back over the sporangial clusters. This arrangement is sometimes called a false indusium. *Dennstaedia* indusia are cup-shaped. *Polypodium* has naked sporangia, that is, the clusters of sporangia are not covered by any protective tissue.

4. Fern Gametophyte

a. Living

Although fern sporophytes are the dominant generation in a fern life cycle, the small, short-lived fern gametophytes are independent and photosynthetic. Fern gametophytes are common in greenhouses where ferns are grown and typically grow on clay pots or on greenhouse benches. For best viewing, look at a living gametophyte with a dissecting microscope. The small, flat thalli of fern gametophytes are often heart-shaped and are anchored to the substrate by rhizoids extending from the lower surface. Antheridia and archegonia develop on the lower surface. Sperm from antheridia swim to and fertilize the archegonial egg cells, starting the next sporophyte generation.

b. Prepared slides

Obtain a prepared slide of a gametophyte with antheridia and archegonia and a slide with a gametophyte with a young sporophyte. Compare these preparations with the living material. Note how the sporophyte emerges from the region of archegonia and how the sporophyte's young leaves form fiddlenecks.

PART E COMPARISON OF BRYOPHYTES AND LIVING SEEDLESS VASCULAR PLANTS

Table 19-1   Comparison of Bryophytes and Living Seedless Vascular Plants

Division Genus	Bryophytes	Psilophyta	Lycophyta		Sphenophyta	Pterophyta
			Lycopodium	*Selaginella*		
Sporophyte	dependent on gametophyte	dominant	dominant	dominant	dominant	dominant
Stems	no	yes	yes	yes	yes	yes; mostly rhizome
Leaves	no	no	microphylls	microphylls	microphylls	megaphylls
Roots	no	no	yes	yes	yes	yes
Strobili	no	no	in some	yes	yes	no
Spores	homosporous	homosporous	homosporous	heterosporous	homosporous	most are homosporous
Gametophyte	dominant	subterranean	subterranean	inside spore wall	photosynthetic thallus	photosynthetic thallus
Motile sperm	yes	yes	yes	yes	yes	yes

QUESTIONS FOR THOUGHT AND REVIEW

1. What anatomical tissue gave plants such as *Psilotum* the ability to grow upright, branched sporophytes?_____

_____

2. What is the function of spore elaters in *Equisetum*?

_____

3. Describe one morphological feature common to *Equisetum* and *Calamites*. _____

_____

4. In *Selaginella*, the antheridium is enclosed with the

_____.

5. Carboniferous forests were dominated by members of the _____ and

_____ divisions of

seedless vascular plants.

6. What is a sorus, and where would you look for sori? _____

_____

_____

7. What is the function of the annulus in fern sporangia? _____

_____

_____

8. How do fern sperm cells reach fern egg cells?

_____

_____

9. Compare and contrast homospory and heterospory. _____

_____

_____

_____

_____

10. Describe one advantage of heterospory over homospory. _____

_____

_____

Name: _____

Section Number: _____

KINGDOM PLANTAE: SEEDLESS VASCULAR PLANTS

1. List the four divisions of extant seedless vascular plants.

   a. _____

   b. _____

   c. _____

   d. _____

2. Using definitions, contrast homospory and heterospory. _____

   _____

   _____

   _____

3. Which seedless vascular plant division or divisions contain living heterosporous members?

   _____

   _____

4. During the Carboniferous period, which division or divisions of seedless vascular plants formed the dominant vegetation on Earth? _____

   _____

   _____

5. What is a sporangium? _____

   _____

   _____

6. Where would you typically look to find sporangia on a fern plant? Be specific. _____

   _____

   _____

7. What is the function of an annulus in ferns? _____

   _____

   _____

8. Contrast microphylls and megaphylls. _____

   _____

   _____

   _____

EXERCISE 20

KINGDOM PLANTAE: GYMNOSPERMS

OBJECTIVES

1. Identify a gymnosperm specimen to its correct division: Cycadophyta, *Ginkgo*phyta, Pinophyta, or Gnetophyta.

2. Identify selected genera within the Pinophyta.

3. Describe the origin and development of tissues contributing to the structure of a mature pine seed.

3. Understand the stages of the pine life cycle. Name each stage, including its ploidy, and describe the importance of each stage to the overall pine life cycle.

4. Identify the parts of a pine seed cone and a pine seed.

TERMINOLOGY

archegonium	micropyle
dioecious	microsporangium
egg cell	microsporophyll
embryo	monoecious
fascicle	ovulate cone
integument	ovule
intercalary meristem	ovuliferous (cone) scale
megagametophyte	pollen cone
megasporangium	pollen grain
(nucellus)	pollen tube
megasporocyte	seed coat
microgametophyte	

INTRODUCTION TO THE GYMNOSPERMS

The term gymnosperm is an informal designation for a group of plants with seeds but without flowers. There are four divisions with living members: Cycadophyta, Ginkgophyta, Pinophyta (also called Coniferophyta), and Gnetophyta. In addition to bearing seeds, all gymnosperms are woody plants. In the last exercise the heterosporous, but seedless, vascular plant *Selaginella* was studied; today's exercise will reveal that the heterorporous condition is a prerequisite to seed development. All seed plants, including gymnosperms, are heterosporous. A seed consists of a sporophyte embryo plus food reserves all enclosed within a protective **seed coat**. In gymnosperm seeds the female gametophyte serves as food reserve tissue. The seed coat develops from the **integument**, a tissue not present in seedless plants. An integument is a protective tissue that encloses a **megasporangium**. A megasporangium surrounded by an integument is called an **ovule**, and it is the ovule that matures into a seed. Seed plants also produce pollen. Pollen is an immature male gametophyte enclosed within the microspore wall and released from the **microsporangium**. When pollen lands on a receptive ovule, it germinates and develops a **pollen tube**, which delivers sperm cells directly to egg cells. Because pollen delivers sperm cells to egg cells without the need of free water, pollen was a significant development for plants growing on land. The elimination of the need for free water to accomplish fertilization and the origin of the new sporophyte generation enclosed within a protective and nourishing seed led to widespread colonization of dry land by gymnospermous plants.

PART A GYMNOSPERM DIVERSITY

Specimens from the four different gymnosperm divisions are displayed. Examine each specimen and develop a list of characteristics separating the four divisions from one another and distinguishing the three genera within the Gnetophyta.

MATERIALS

- living cycad
- branch cut from *Ginkgo* tree
- prepared slide of *Ginkgo* leaf, whole mount
- living *Ephedra*
- branch cut from *Gnetum* vine
- living *Welwitschia*
- branches cut from pine, redwood, juniper, cypress, and other available conifers
- a conifer key
- prepared slide of *Pinus* leaf in cross section
- prepared slide of *Pinus* wood in cross, tangential, and radial sections
- dissecting microscope
- compound microscope

PROCEDURE

1. Division Cycadophyta

   Cycads were dominant land plants during the Jurassic period—the same period in geological history when dinosaurs reigned. To zoologists the Jurassic was the "Age of Dinosaurs"; to botanists it was the "Age of Cycads." Most cycads are extinct. However, there are nine or ten surviving genera, which are restricted to tropic and subtropic environments. Economically, the cycads are valuable as landscape specimens. They are usually unbranched with a stout, thick stem and large pinnately compound leaves and are often mistaken for short palm trees. Plants have a tap root and adventitious roots from the base of the stem. Secondary growth is not extensive; stems have large pith and cortex regions. All cycads are **dioecious** (they have separate male and female plants) with large strobili. Because cycads are seed plants, they have pollen, but they also retain a relict feature—swimming sperm. However, the sperm swim only within the pollen tube. Cycad pollen is dispersed by wind or insects. Examine the specimens on display and list morphological characteristics to help in distinguishing cycads from other plants.

2. Division Ginkgophyta

   There is only one living species in this division, *Ginkgo biloba*. *Ginkgo* is a large, highly branched tree with beautiful fan-shaped leaves that are winter deciduous, turning golden yellow in the fall. The division had a worldwide distribution during the Triassic and Jurassic periods, and today *Ginkgo biloba* is widely cultivated as a landscape tree. There has been little change in this plant in almost 300 million years of existence, and it is sometimes called a living fossil. *Ginkgo* has an active vascular cambium that produces much secondary growth. There are two types of branches, long shoots and short shoots. The short shoots, also called spur shoots, bear leaves. The species is dioecious with small strobili borne on short shoots. It is recommended that male plants be used in landscaping because mature *Ginkgo* seeds, which, of course, are borne on female plants, contain butryic acid, which produces an objectional smell. Examine the potted specimen or branch on display. Identify long and short shoots and note leaf shape and venation. Use a dissecting microscope to examine the prepared slide of a whole mount of a *Ginkgo* leaf and note the dichotomous leaf venation.

3. Division Pinophyta

   This division contains plants informally known as conifers. Most conifers are trees, some are shrubs, and the majority are evergreen. They form the dominant vegetation in cool-temperate regions of both hemispheres. Many are adapted to xeric environments and some, such as the bald cypress, *Taxodium*, grow in swamps. Conifers form extensive forests in western North America. This division has great economic value; conifers are harvested for lumber, paper, fuel, turpentine, and resins, and at least one, the Pacific yew, has gained notice recently because of its production of the anticancer compound taxol. Conifers are also important in landscaping.

   The xylem contains tracheids, but no vessels, and the phloem sieve cells lack companion cells. Conifers have extensive secondary growth produced from a vascular cambium, and the wood produces distinct annual growth rings. Wood from conifers is called softwood, although it may be harder than some angiosperm wood, which is known as hardwood.

   a. Conifer keying exercise

      There are seven families with over 700 species within the division Pinophyta. The most familiar family to North Americans is the Pinaceae, which includes pines, firs, spruces, hemlocks, Douglas firs, and larches. Redwoods are in the Taxodiaceae family, and junipers and cypresses are in the Cupressaceae family. A dichotomous key will be provided that was designed for the Pinophyta specimens available. (A dichotomous key is a tool used to identify specimens to a correct taxon level.) Traits used to distinguish the genera include leaf shape, leaf arrangment along the stem, whether leaves are borne singly or in

fascicles, whether plants are monoecious or dioecious, and whether or not ovules are borne in cones. Use the key to determine the correct genus for each of the specimens displayed.

b. Pine leaf anatomy

Pine leaves demonstrate adaptations to xeric conditions. Examine a pine leaf cross section with a compound microscope. Note the thick cuticle, thick-walled hypodermal cells, and stomata sunken in pits. Look for the distinctive resin ducts within the leaf mesophyll and the two veins enclosed within an endodermis. Between the veins and the endodermis is a region, called transfusion tissue, filled with additional tracheids.

c. Pine wood

Obtain a prepared slide of pine wood and examine it with a compound microscope. First examine the cross section. Note the uniformity of the cells, the presence of resin ducts, and the growth rings. Now examine the radial and longitudinal sections. Again note the uniformity of cells when compared to the complex wood of oak seen in the exercise on secondary growth in angiosperms.

4. Division Gnetophyta

There are three highly specialized genera within this division: *Gnetum*, *Ephedra*, and *Welwitschia*. These plants are more like angiosperms than any other gymnosperm. They have secondary xylem with vessels and tracheids and compound strobili that resemble flowers.

Gnetum inhabits tropical forests in the Amazon Basin of South America, in Africa, and in Asia. *Gnetum* is a vine, shrub, or tree, and looks much like an angiosperm.

Ephedra is a xerophytic plant with a wide distribution in tropical and temperate North America, South America, and Asia. It somewhat resembles a shrubby *Psilotum* or *Equisetum*; however, it is a woody plant. The stems are green, the leaves are scalelike, and the lateral branches occur in whorls.

Welwitschia is xerophytic, but it is restricted to a narrow coastal fog desert in southwestern Africa. *Welwitschia* has an enormous tap root, a very broad, short stem, and only two adult leaves. The leaves grow at the base from an **intercalary meristem** and persist for the life of the plant, splitting at the ends and giving the appearance of more than two leaves. Examine the specimens available and note any traits mentioned above that are easily visible.

The genus *Pinus* will be used as a representative organism to study the development of seeds and pollen within the gymnosperms. As seen earlier, the heterosporous seedless vascular plant *Selaginella* bears megasporangia exposed on the surface of sporophylls, but in seed plants a layer of tissue encloses and protects megasporangia. The tissue surrounding a megasporangium is called an integument. A megasporangium surrounded by an integument is an ovule, and it is the ovule that becomes a seed. As the embryo develops, the integument differentiates into the seed coat. Seeds have three major parts: an embryo plant, food reserves for the embryo, and a protective seed coat. Before examining the pine reproductive specimens and slides, study the pine life cycle diagram (Figure 20-1, page 162).

MATERIALS

- living or dried male *Pinus* cones containing pollen grains
- variety of female pine cones with seeds
- prepared slide of a mature *Pinus* male cone with pollen
- prepared slide of a young female *Pinus* cone at the megasporocyte stage
- prepared slide of female *Pinus* cone with **mature megagametophytes**
- fresh pine seeds
- prepared slide of mature pine seed
- microscope slides
- coverslips
- dH$_2$O in dropper bottle
- single-edge razor blade
- compound microscope

PROCEDURE

1. The Male Strobilus

a. Fresh pollen cones

Other terms for male strobili include **pollen cones**, male cones, or pollen strobili. When mature, individual pollen cones have a papery texture and are smaller than the woody seed cones. When pollen cones are mature, **pollen grains** fall from the cones in clouds of yellow dust. If available, examine the fresh specimens of pine branches with attached male strobili, or examine the preserved cones. Starting from the bottom of a fresh cone, remove **microsporophylls** and determine the pattern of sporophyll arrangement around the cone axis. Use a dis-

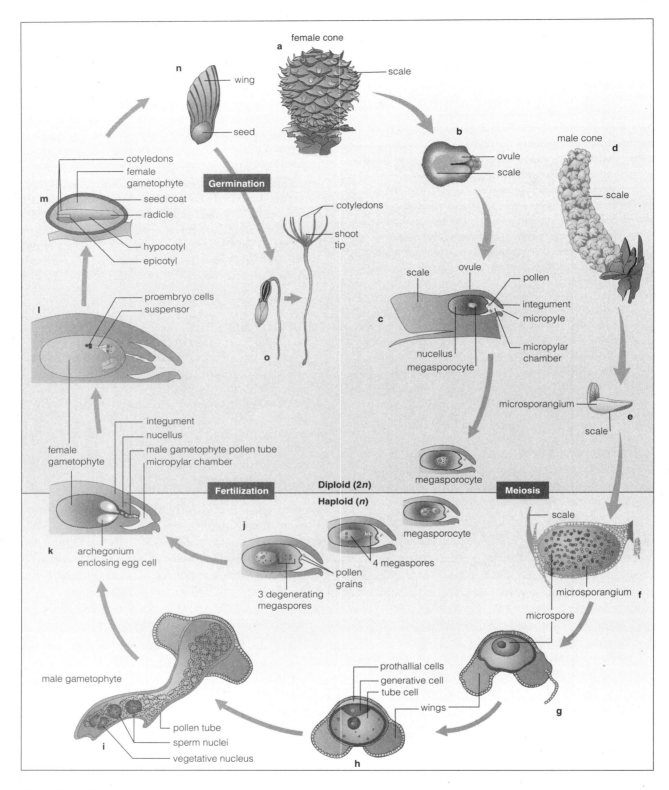

Figure 20-1 Pine life cycle stages.

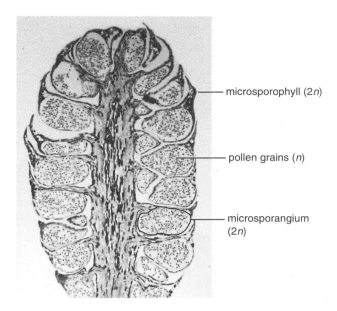

Figure 20-2 Pine male strobilus with pollen grains, longitudinal section.

microsporophyll (2*n*)

pollen grains (*n*)

microsporangium (2*n*)

secting microscope to examine a single microsporophyll. Note which side bears microsporangia and the number of sporangia per sporophyll. Break open a sporangium to release pollen grains, and make a wet mount of the pollen. Examine the pollen preparation with a compound microscope. Each pine pollen grain has two wings; these are extensions of the cell wall and aid in directing the growth of the pollen tube toward the egg cell.

b. Prepared slide of pollen cone

After examining intact pollen cones, pick up a prepared slide of mature male cones containing pollen. Use a compound microscope to study the anatomical structure. First, examine the slide at the lowest magnification and determine if the slide contains a single male cone or a cluster of male cones; concentrate on a single cone (Figure 20-2).

Confirm the arrangement of sporophylls around the cone axis and the location of individual microsporangia on the underside of sporophylls. Change to a higher magnification and look for individual pollen grains. Note the pollen grain wings and study the individual cells and nuclei of the pollen grain. A mature pine pollen grain contains four cells: two prothallial cells, a generative cell, and a tube cell. Pine pollen is wind dispersed. After pollen grains land on the female cone and the grains germinate, the pollen tube nucleus directs the growth of the pollen tube, and the generative cell divides to produce sperm cells. Once the

pollen tube and sperm cells are present, the microgametophyte is mature.

2. The Female Strobilus

a. Ovulate cone diversity

Examine the diversity of sizes and shapes of the female cones on display. These are the traditional "pine cones." Female cones are also called **ovulate cones**, seed cones, or female strobili. Unlike male cones, female pine cones are compound structures. The ovule-bearing structures that attach to and spiral around the cone axis are modified branches called cone scales or **ovuliferous scales**. Below each cone scale is a papery leaf called a bract. The length and complexity of the bract varies with the species. Each cone scale bears two ovules and each ovule produces one seed. Pollination occurs when the ovulate cone is small and the cone scales are thin and well separated from one another. After pollination, the cone increases in size and becomes woody.

b. Prepared slide of ovulate cone at megasporocyte stage

Obtain a prepared slide of an ovulate cone in longitudinal section; and, with the aid of a compound microscope, locate an intact cone scale bearing an ovule (Figure 20-3).

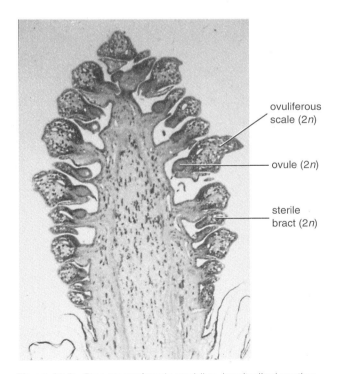

ovuliferous scale (2*n*)

ovule (2*n*)

sterile bract (2*n*)

Figure 20-3 Pine young female strobilus, longitudinal section.

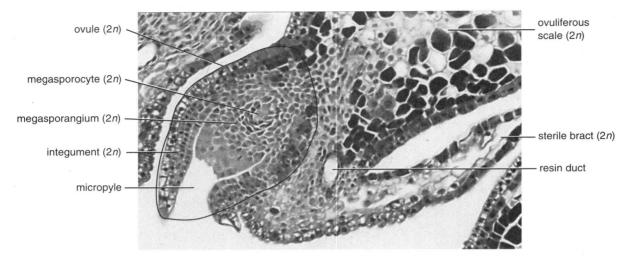

Figure 20-4 Pine ovuliferous scale at megasporocyte stage, longitudinal section.

The entire ovule should be visible (Figure 20-4), including the integument with its small opening, the **micropyle**, and the enclosed megasporangium with its large **megasporocyte** (megaspore mother cell). (In gymnosperms the megasporangium is also called a **nucellus**.) Pollination occurs at this stage in pine ovule development.

c. Prepared slide of ovulate cone at megagametophyte stage

After pollination and following months of ovule development, the **megagametophyte** reaches maturity with its **egg cells** ready for fertilization. Obtain a slide with a mature pine megagametophyte (Figure 20-5), and starting with the outermost tissue of the ovule, identify the integument, the remains of the megasporangium, and the mature megagametophyte with its **archegonia**. Each archegonium contains a huge egg cell. Egg cell nuclei are larger than the sterile jacket cells of the archegonium.

Observe the demonstration slide of a mature pine seed in longitudinal section. The seed coat has been removed. Two tissues are visible: the megagametophyte and the **embryo**. The megagametophyte serves as food reserves for the embryo, which is the next sporophyte generation.

3. The Pine Seed

a. Fresh seed

If available, remove a fresh seed from the scale of a mature seed cone. Note the overall appearance of the seed including the hardness of the seed coat and the presence of a thin wing at one end of the seed. Large pine seeds

(sometimes erroneously called pine nuts) are available for dissecting. Using your fingernails, remove the seed coat and any remaining brown tissue from the megasporangium. The exposed white tissue is a megagametophyte, the tissue that serves a nutritive function for the embryo. With a fresh razor blade, make a shallow cut around the megagametophyte and pull it away to expose the embryo. Pine embryos have many thin cotyledons and a short radicle.

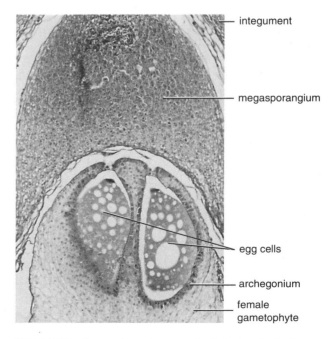

Figure 20-5 Pine mature megagametophyte, longitudinal section.

b. Prepared slide of a pine embryo

Use a compound microscope at low power to study the anatomy of a pine embryo sectioned and mounted on a prepared slide. The preparation reveals procambium tissue within the embryo axis. Note the surrounding megagametophyte tissue. The seed coat was removed prior to sectioning the tissue for microscope preparation.

QUESTIONS FOR THOUGHT AND REVIEW

1. Name and describe the type of leaf venation present in *Ginkgo* trees.

_____

_____

2. Which gymnosperm division has plants with large, frondlike leaves? _____

_____

3. Which genus in the Gnetophyta can be found in western U.S. deserts?_____

_____

4. How many leaves are present in a mature *Welwitschia* plant?_____

5. How many cells are in a pine pollen grain?

_____

6. Describe two features that distinguish pine pollen grains from mature pine microgametophytes. _____

_____

_____

_____

7. During the pine life cycle, at what stage of ovule development does pollination occur? _____

_____

_____

8. At what stage of ovule development does fertilization occur? _____

_____

_____

9. How many seeds does each pine ovuliferous scale produce? _____

10. Are male and female pine strobili produced on the same tree or on separate male and female trees?

_____

EXERCISE 20
LABORATORY QUIZ

Name: _____

Section Number: _____

KINGDOM PLANTAE: GYMNOSPERMS

1. Why are nonflowering seed plants informally called gymnosperms? _____
_____

2. List the four divisions of extant gymnosperms.

    a. _____

    b. _____

    c. _____

    d. _____

3. Define an ovule. _____
_____
_____

4. What is a seed? _____
_____
_____

5. Are gymnosperms homosporous or heterosporous? _____
Explain. _____
_____

6. How does pine pollen reach an ovulate cone? _____
_____
_____

7. In pine seeds, what tissue serves as nutritive tissue for the embryo, and what is its ploidy? _____
_____
_____
_____

8. What is the origin of the seed coat in pine seeds? _____
_____
_____
_____

EXERCISE 21

KINGDOM PLANTAE: ANGIOSPERMS

OBJECTIVES

1. Summarize the unique traits of the angiosperms.

2. Compare flowering plants relative to traits considered primitive and traits considered advanced.

3. Identify specimens from several representative angiosperm families.

4. Describe several economically important angiosperm families.

TERMINOLOGY

advanced trait   Magnoliopsida
Alismataceae   monocot
dicot   Orchidaceae
family   primitive trait
Lamiaceae   taxon
Liliopsida   tepal
Magnoliaceae   whorl
Magnoliophyta

INTRODUCTION TO THE ANGIOSPERMS

We began this course with a study of flowering plant structure and function, and then briefly examined significant traits of the other walled organisms—bacteria, fungi, and algae. We returned to our study of plants by examining new evolutionary traits invented by increasingly complex members of the kingdom Plantae, including bryophytes, seedless vascular plants, and gymnospermous seed plants. These plant divisions exhibit trends in better adaptation to life on land and an increasingly complex sporophyte generation. Refer to Table 21-1, page 170, for comparison of some of the important traits of heterosporous seedless vascular plants, gymnosperms, and angiosperms.

In this exercise, we return to the only **taxon** of plants that produce flowers, the division Magnoliophyta. You have already seen that the flowering plants have ovules that are borne in a container, the

ovary, that matures into a fruit. The ovary and the fruit are unique characteristics of the angiosperms. Double fertilization that leads to the formation of endosperm is another invention of the angiosperms. Today you will examine this most advanced group of land plants in light of evolutionary relationships among some of flowering plant taxa. In addition, you will examine some of the economically important families and genera of angiosperms.

PART A PRIMITIVE AND ADVANCED FLOWER FAMILIES

The dicots, **Magnoliopsida**, and the monocots, **Liliopsida**, apparently diverged early in the evolution of the **Magnoliophyta**. Because the flower is central to determination of flowering plant species, the flower is the most important source of information in determining evolutionary relationships among taxa of angiosperms. In Exercise 10, we examined the basic structure of flowers and saw that flower structure is amazingly diverse. It will be helpful to review some of the terminology from Exercise 10 before proceeding.

Recall that much of the variation in flowers reflects the following differences among the **whorls** of floral organs: (1) increase or decrease in the number of parts in each whorl, (2) fusion of parts within or between whorls, (3) changes in the symmetry of a flower, (4) apparent position of the ovary relative to the other whorls, and (5) presence or absence of specific whorls.

In addition, you saw that the flowers of species within the Magnoliopsida generally have floral parts in fours or fives or multiples of fours or fives, such as five sepals and ten stamens, and that the flowers of species within the Liliopsida typically have floral parts in groups of three or multiples of three. Other potentially important features of flowers include fusion of floral organs, particularly when resulting in a hypanthium; actinomorphic (or regular) symmetry versus zygomorphic (or irregular) symmetry; and superior versus inferior ovaries.

Generally, trends in flower structure are associated with the evolution of flowering plants. Those traits that appeared early in the evolution of the angiosperms

Table 21-1 Comparison of *Selaginella*, Gymnosperms, and Angiosperms

Informal Group Name	SVP		Gymnosperms			Angiosperms
Division	Lycophyta	Cycadophyta	Ginkgophyta	Coniferophyta	Gnetophyta	Magnolio-phyta
Genus	*Selaginella*					
Sporophyte	dominant	dominant	dominant	dominant	dominant	dominant
Spores	heterosporous	heterosporous	heterosporous	heterosporous	heterosporous	heterosporous
Gametophyte	endosporic*	endosporic*	endosporic*	endosporic*	endosporic*	endosporic*
Archegonia	yes	yes	yes	yes	*Ephedra*—yes *Welwitschia*—no *Gnetum*—no	no
Antheridia	yes	no	no	no	no	no
Motile sperm	yes	yes	yes	no	no	no
Pollen	no	yes	yes	yes	yes	yes
Ovules and seeds	no	yes	yes	yes	yes	yes
Double fertilization	no	no	no	no	*Ephedra*—yes *Welwitschia*—? *Gnetum*—yes	yes
Endosperm	no	no	no	no	no	yes
Ovary wall and fruit	no	no	no	no	no	yes

*endosporic = micro/megagametophyte retained within micro/megaspore wall

are called **primitive traits** and those that appeared later are called **advanced traits**. The terms primitive and advanced imply nothing about the quality of these traits.

Trait	Primitive Condition	Advanced Condition
Number of flower parts	Many, indefinite in number	Few, definite in number
Number of whorls	Four	Three, two, or one
Arrangement of parts	Obviously spiral	Spiral not evident
Fusion of floral parts	No	Yes
Ovary position	Superior	Inferior
Floral symmetry	Regular	Irregular

In this part of the exercise, you will compare a primitive and an advanced dicot family and a primitive and an advanced monocot family. Refer to the traits listed above as you examine the representative flowers.

MATERIALS

- *Magnolia* flower
- mint flower
- water-plantain flower
- orchid flower
- dissecting microscope

PROCEDURE

1. Primitive and Advanced Families of the Magnoliopsida

   a. Magnoliaceae—a primitive dicot family

   The plants in the magnolia **family** are either trees or shrubs with simple leaves and solitary flowers. Within North America, two common members of this family are *Magnolia* and *Liriodendron*. Examine a *Magnolia* flower. Can you distinguish between sepals and petals? In flowers with similar-appearing sepals and petals, the organs of the perianth are called **tepals**. Particularly note the elongated receptacle and distinct spiral arrangement of the stamens and carpels. *Magnolia* flowers are fragrant and attract beetles for pollination. Complete the following:

Species: _____

Part or feature	Number of parts
sepals	_____
petals	_____
stamens	_____
carpels	_____
locules	_____

Characteristic

fusion of similar parts	_____
fusion of dissimilar parts	_____
symmetry	_____
ovary position	_____
complete or incomplete	_____
perfect or imperfect	_____

b. Lamiaceae—an advanced dicot family

The mint family, also called the Labiatae, includes annuals, perennial herbs, shrubs, and trees. The stems are often square in cross section and many contain fragrant oils. Leaves range from simple to compound and flowers are borne in various types of infloresences. Examine the mint flower provided, and complete the following:

Species: _____

Part or feature	Number of parts
sepals	_____
petals	_____
stamens	_____
carpels	_____
locules	_____

Characteristic

fusion of similar parts	_____
fusion of dissimilar parts	_____
symmetry	_____
ovary position	_____
complete or incomplete	_____
perfect or imperfect	_____

2. Primitive and Advanced Families of the Liliopsida

a. Alismataceae—a primitive monocot family

The Alismataceae is commonly called the arrowhead or water-plantain family. These are annual or perennial aquatic herbaceous plants. They produce a milky sap and sometimes have tubers. The leaves are simple with long petioles. Obtain a water-plantain flower, and determine why this family is considered primitive. Complete the following:

Species: _____

Part or feature	Number of parts
sepals	_____
petals	_____
stamens	_____
carpels	_____
locules	_____

Characteristic

fusion of similar parts	_____
fusion of dissimilar parts	_____
symmetry	_____
ovary position	_____
complete or incomplete	_____
perfect or imperfect	_____

b. Orchidaceae—an advanced monocot family

The orchid family is a familiar group of showy plants widely distributed throughout the world, but they are at their most diverse in the tropics. It is this family from which vanilla is obtained. Most orchids are terrestrial or epiphytic perennial herbs. Observe the orchid flowers provided and particularly note the distinctive zygomorphic symmetry of this taxon. Complete the following:

Species: _____

Part or feature	Number of parts
sepals	_____
petals	_____
stamens	_____
carpels	_____
locules	_____

Characteristic

fusion of similar parts _____

fusion of dissimilar parts _____

symmetry _____

ovary position _____

complete or incomplete _____

perfect or imperfect _____

- safflower oil
- artichoke
- sunflower seeds
- lettuce
- oats
- barley
- rice
- wheat
- corn
- sugarcane

PART B SOME ECONOMICALLY IMPORTANT PLANTS

Although the gymnosperms are an important source of lumber products, it is the angiosperms that provide our most important economic plants. Angiosperms provide lumber, food, clothing, cordage, and clean air. All true fruits come from flowering plants, and the vegetative plant organs, leaves, stems, and roots provide much of our food and fiber. The grass family alone provides cereal products such as rice, corn, wheat, and millet; and the legume family gives us peanuts, peas, beans, lentils, alfalfa, and clover. Cotton and flax are examples of plants from which we obtain fibers for making cloth. In this portion of the exercise, concentrate on the plant organ from which products are derived and the family of the various plant products.

MATERIALS

- strawberry fruit
- apple fruit
- cherry fruit
- pear fruit
- blackberry fruit
- rose hips
- horseradish
- broccoli
- cabbage
- cauliflower
- Canola oil
- radish seedlings
- mustard seeds
- peanuts
- soy beans
- beans
- peas
- clover
- lentils

PROCEDURE

Examine plant products and the representative flowers from each of the families on display. The representative families contain many important commercial crops. Determine from which part of the plant each product is derived. Look for similarities within products from the same family. Fill in the following information as you proceed through the display.

Family: Rosaceae, the rose family

Species of representative flower:_____

Product	Plant organ
_____	_____
_____	_____
_____	_____
_____	_____
_____	_____
_____	_____

Family: Brassicaceae, the mustard family

Species of representative flower:_____

Product	Plant organ
_____	_____
_____	_____
_____	_____
_____	_____
_____	_____
_____	_____

Family: Fabaceae (Leguminosae), the legume family

Species of representative flower:_____

Product	Plant organ
_____	_____
_____	_____
_____	_____
_____	_____
_____	_____
_____	_____

Family: Asteraceae (Compositae), the sunflower family

Species of representative flower:_____

Product	Plant organ
_____	_____
_____	_____
_____	_____
_____	_____
_____	_____
_____	_____

Family: Poaceae (Gramineae), the grass family

Species of representative flower:_____

Product	Plant organ
_____	_____
_____	_____
_____	_____
_____	_____
_____	_____
_____	_____

QUESTIONS FOR THOUGHT AND REVIEW

1. List two important traits of angiosperms not found in other plant taxa.

  a._____

  b._____

2. Compare the floral symmetry of plants in the Magnoliaceae with plants in the Orchidaceae.

_____

_____

_____

3. Describe one major difference between dicots and monocots. _____

_____

4. An inferior ovary is more advanced than a superior ovary. Of what advantage might an inferior ovary be to a plant? _____

_____

_____

5. Plant family names end in what five letters?

_____

6. What is wrong with this statement: "Primitive traits are of poor quality compared to advanced traits." _____

_____

7. What type of symmetry does an actinomorphic flower have? _____

_____

8. List the floral organs most likely present in a primitive flower._____

_____

_____

9. Describe one floral characteristic of the Lamiaceae that is an advanced trait. _____

_____

_____

10. List five commercially valuable products obtained from the Brassicaceae.

a. _____

b. _____

c. _____

d. _____

e. _____

EXERCISE 21
LABORATORY QUIZ

Name:_____

Section Number: _____

KINGDOM PLANTAE: ANGIOSPERMS

1. List two defining features of angiosperms.

    a. _____

    b._____

2. Describe a typical difference between dicot and monocot flowers. _____

_____

_____

3. If a given flower has dozens of stamens, a superior ovary, and regular symmetry, would you expect the plant to be classified in a primitive or an advanced family? _____

_____

4. Are orchids monocots or dicots?_____

5. Peas and beans are classified in which plant family? _____

6. Rice, wheat, and sugarcane are all members of which plant family? _____

_____

7. Are sunflowers and orchids closely related? Why or why not? _____

_____

_____

8. Where would you look for a female gametophyte in a flowering plant?_____

_____

_____

_____

EXERCISE 22

COMPETITION EXPERIMENT

OBJECTIVES

1. Describe and compare the growth of unitary and modular organisms.

2. Describe the results and their significance of an intraspecifc competition experiment involving modular organisms.

TERMINOLOGY

biomass	normal distribution
density	population
intraspecific competition	rosette plant
modular organism	unitary organism

INTRODUCTION TO A COMPETITION EXPERIMENT

Populations of **unitary organisms** are composed of individuals that are genetically unique, physiologically independent of one another, and constructed of nonrepetitive body parts. Most animals, such as mammals, are unitary organisms. To follow the growth of a population of unitary organisms we would usually count the number of individuals in a specified area over a period of time. For convenience, counts may be made of only those individuals in a single life cycle stage: for example, adults. Merely by counting individuals, much can be learned about a population, and some predictions can be made about the population's future, at least over the short term.

All plants and some animals, such as certain marine invertebrates, are **modular organisms**. Modular organisms are composed of repeating units of individual parts or modules. In many species the modules can become physiologically independent from one another, and modular construction of organisms allows for considerable variation in individual size and architectural flexibility. For example, when the environment is favorable, a modular organism can grow new parts; when conditions are less favorable, parts can be shed. The modular nature of plants requires a different method of measuring the growth of a population, and measuring **biomass** instead of counting is usually the preferred method.

Not only do modular organisms show greater individual size variation than unitary organisms, but the distribution of size also differs from that of unitary organisms. Because body size is genetically determined in unitary organisms, we typically find a **normal** (or bell-shaped) **distribution** for a population of unitary organisms. In contrast, most populations of modular organisms show a distinctly L-shaped distribution (Figure 22-1, page 178).

In this exercise, you will learn some of the ways that resource availability affects the individual growth and the population growth of a plant species. **Intraspecific competition**, or competition between members of the same species, will be used to explore the affects of resource availability. Using radish plants as an example of a modular organism, a series of radish populations will be established at different **densities**, and the growth of the radish populations will be compared at the end of a specified length of time.

Radish (*Raphanus sativus*) is an important root vegetable crop in the family Brassicaceae. Radish seeds are relatively large (2–3 mm in diameter) and germinate readily once watered. Within a week, the paired cotyledons of seedlings are fully expanded, and within another week, the first true leaves are formed. Growth proceeds in a modular fashion. New leaves are formed throughout the juvenile stage, giving rise to a larger and larger **rosette**. Within a few months, the rosette switches from vegetative growth to sexual reproduction. New leaves cease being formed, and a flowering shoot begins to emerge from the center of the rosette. Most radish cultivars have been selected to delay the onset of flowering, so it is unlikely that any of your plants will reach reproductive maturity over the course of the academic term.

PART A   EXPERIMENT SETUP

Today, the experiment will be set up by establishing radish populations in a series of finite and repeatable growth environments using different numbers (densi-

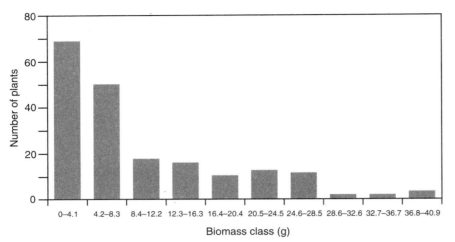

Figure 22-1 The size distribution of individual plant biomass in a population of *Raphanus sativus*, shown as the number of plants in 10 biomass classes. The biomass classes were determined by dividing the biomass range into 10 equal intervals. n = 170.

ties) of individuals. Pots of soil will be planted with different densities of seeds, and the growth of the individuals and the growth of the population in each pot will be monitored. Near the end of the term, you will count radish plants and determine the biomass of individual radish plants and the biomass of radish populations from each experimental population.

Your instructor will assign your laboratory class to five different groups. Each group will be responsible for planting an entire population density series of radish plants, gathering the growth data, and sharing the data with the other four groups in the laboratory class.

MATERIALS

- 4 10 cm × 10 cm greenhouse pots
- greenhouse soil to fill pots
- 30 seeds of Scarlet Globe radish
- 4 labels
- 1 Sharpie® permanent marker, black
- greenhouse space

PROCEDURE

1. Each group will set up an entire density series of populations of radish plants: 2, 4, 8, and 16 individual plants (seeds). Note that the density series doubles at each treatment level.

2. Use a Sharpie® pen and label four greenhouse pots exactly as described below. (Warning: *Do not* use ink or pencil to make the labels; ink will wash away and pencil is too difficult to read after weeks in the greenhouse. If you do not have a Sharpie® pen, borrow one.)

Group letter or code (assigned by your instructor)

Day of week, lab meeting time, and room number

Density treatment (the number of seeds in the pot)

For example:

Group G

Mon., 9-12, Rm 183

4 seeds

3. Fill the pots with potting soil so that loose soil is exactly even with the top of the pots. Using an empty pot, lightly tamp down the soil so that the surface is 0.5 cm below the top of each pot. Take the pots to a sink and water the soil. Leave the pots to drain completely. (This should take no more than 5 minutes.)

4. *Important! Read this entire paragraph before planting.* After the pots have drained, place 2, 4, 8, or 16 seeds on the soil in each pot, ensuring that the number of seeds per pot corresponds to the pot label. You should have one pot with 2 seeds, one pot with 4 seeds, one pot with 8 seeds, and one pot with 16 seeds. *DO NOT put extra seeds in any of the pots.* To do so would destroy the integrity of the experiment. (Why?) Avoid seeds with shrivelled or cracked seed coats. Distribute the seeds as evenly as possible onto the surface of the wet soil. Gently press each seed into the soil surface, but do not bury the seeds. Now, add about about 0.25 cm of soil, covering the seeds. Do not water. Your instructor will show you where to place your pots for later transfer to the greenhouse.

5. Record your group code: _____

List the names of fellow students in your group:

_____

_____

_____

_____

_____

PART B DATA COLLECTION

After the designated length of growing time, the plants will be harvested. In each population, count individuals and then weigh them collectively to measure total biomass. For the high-density treatment (16 starter seeds), weigh individual plants to generate a biomass distribution for plants growing under a competitive regime. Decide how to efficiently partition tasks among group members. Recall that the top-loader balance scale will be heavily used, so please coordinate your activities with other groups.

MATERIALS

- pots of radishes from greenhouse
- scissors
- small paper bags
- top-loader balance scale

PROCEDURE

1. For this part of the laboratory exercise, work with a piece of newspaper under the pots to avoid getting soil on the lab bench and on your data sheets. Also, because the pots and soil may be wet, record your data in pencil or in indelible ink. Begin your data collection with the pot for the 2-seed density treatment, and follow the steps described below.

    a. In Table 22-1 record the number of plants growing in the pot. Each individual has a single stem, with leaves emerging in a rosette from that stem. If leaves are so dense that you cannot reliably count individual plants, then keep a "running total" as you harvest individual plants.

    b. Working with sharp scissors or a single-edged razor blade, cut off plants one at a time at the base of the green shoot, do not include the red storage organ. Weigh the plants on the top-loading balance. To do this, place a paper bag on the balance and tare. Put all of the plants from a single pot in the bag and record the weight in Table 22-1.

2. Repeat the above steps for the 4- and 8-seed density pots.

3. For the 16-seed density treatment, follow the same steps as above, with one important modification. In this treatment, you will be recording the weights of individual plants. Record the weight of each plant in Table 22-2. In Table 22-1, record the number of plants and total plant weight from this treatment.

Table 22-2   Biomass of Individual Shoots in the Group 16-Seed Density Treatment

Plant number	Plant biomass (g)	Plant number	Plant biomass (g)
1		9	
2		10	
3		11	
4		12	
5		13	
6		14	
7		15	
8		16	

Table 22-1   Radish Plant Population Group Data

Density treatment	Total number of plants per treatment (pot)	Total shoot biomass (g) per treatment (pot)
2 seeds		
4 seeds		
8 seeds		
16 seeds*		

*Note: Plants in 16-seed density treatment are weighed and measured individually.

4. Pool the radish data with the rest of your lab section, recording all data in Tables 22-3, 22-4 and 22-5.

Table 22-3   Total Number of Radish Plants

Group number	Density treatment			
	2 seeds	4 seeds	8 seeds	16 seeds
1				
2				
3				
4				
5				
Mean				
Standard deviation				
Standard error				

Table 22-4   Total Radish Shoot Biomass (g)

Group number	Density treatment			
	2 seeds	4 seeds	8 seeds	16 seeds
1				
2				
3				
4				
5				
Mean				
Standard deviation				
Standard error				

Table 22-5   Individual Radish Shoot Biomass from the 16-Seed Treatment
(In the space allotted for your group, legibly record each shoot biomass, separated by commas, for all the shoots in your 16-seed density treatment. For example: 1.0, 2.5, 3.0, . . .)

Group number	Individual radish shoot biomass (g)
1	
2	
3	
4	
5	

PART C PREPARING A REPORT OF THE EXPERIMENT

In your own words, prepare a formal written report of the competition experiment. Use the format described below, and turn in the report to your instructor on the designated date.

PROCEDURE

1. Write an appropriate title.

2. Write an introduction section. In just a few sentences, introduce the report topic and describe the major objectives of the competition experiment.

3. Write a materials and methods section. If you followed the protocols outlined in the lab manual exactly, just make a statement to that effect. If you deviated from lab manual procedures, briefly describe how and why this was done. Do not repeat the protocol, but do cite the lab manual.

4. Prepare and write a results section. In the results section of your written report, briefly describe what you observed. Save the interpretation of why you got these results for the discussion. The results is where you present the findings of your experiment. Your results section will include graphs, tables, and written descriptions, based on radish data collected from all five groups within your lab section. You will be preparing five tables and five graphs; the templates for the graphs are on page 183. Graphically summarize your data as outlined below, and turn in the tables and graphs with the report. Incorporate answers to the questions posed below in the written portion of the results section.

   a. Calculate the means and standard deviations for the total number of radish plants per density treatment (pot), and record the values in Table 22-3. Prepare a line graph (Figure 22-2, page 183) showing the means and including standard error bars.

   b. Calculate the means and standard deviations for the total shoot biomass per density treatment, and record the values in Table 22-4. Then line graph (Figure 22-3, page 183) the means with standard error bars.

   c. In Table 22-5, page 180, record the individual biomass values from each group's 16-seed density treatment.

   d. Record the biomass classes in Table 22-6, page 182. Now, tally the number of plants from Table 22-5 that fall within each biomass class

and record the tally marks in the spaces provided in Table 22-6. Count and record the total number of plants for each biomass class.

   e. Prepare a histogram (Figure 22-4, page 183) showing the distribution of individual plant biomass sizes. Label the horizontal axis with the calculated biomass categories.

   f. Complete Table 22-7, page 182, by determining the mean shoot biomass per plant and the percent plant survivorship, then line graph the data (Figures 22-5 and 22-6, page 183).

   g. Incorporate the following questions into the results section:

      i. What relationship, if any, did you observe between the starting density and the final population density?

      ii. What was the relationship between starting density and the final population biomass?

      iii. Did the mean shoot biomass per plant remain the same over the four density treatments?

      iv. What happened to plant (seed) survivorship as seed density increased?

5. Write a discussion section. Describe the significance of the results and present your interpretation of the results. Explain what the results mean or why they exist. Use this section to discuss the implications of the results. Frame your final written discussion around the following questions:

   a. Based on the experiment results, what insight have you gained about the nature of population regulation in radish plants?

   b. How would you define carrying capacity for radish plants in this experiment? See the textbook glossary or ask your instructor for a definition of carrying capacity.

   c. As carrying capacity was approached at increasing starting densities, how important were changes in the number of individuals versus changes in the size of individuals?

   d. What factors might give rise to the L-shaped size distribution of the radish population?

   e. Relate your results to broader concepts, such as determinate versus indeterminate growth.

6. Prepare a literature cited section. Alphabetically, list all literature cited in the body of your report. Follow the format suggested by your instructor.

Identifying ten biomass classes for radish plants:

Minimum biomass = _____

Maximum biomass = _____

Maximum biomass − minimum biomass = _____ (= range of biomasses)

Range/10 = _____ (= biomass range increment, I)

Minimum biomass is low end of smallest biomass class = _____

Minimum biomass + I defines low end of second biomass class = _____

Minimum biomass + I defines low end of third biomass class = _____

Continue to determine remaining biomass classes.

Maximum biomass defines high end of largest biomass class = _____

Table 22-6   Distribution of Radish Biomass (g) from the 16-Seed Density Treatment

Class number	Shoot biomass class	Tally counts for this biomass class	Count total
1			
2			
3			
4			
5			
6			
7			
8			
9			
10			

Table 22-7   Mean Shoot Biomass (g) per Plant and Survivorship per Treatment

Density treatment = A	Mean number of plants per treatment = B (from Table 22-3)	Mean shoot biomass per treatment = C (from Table 22-4)	Mean shoot biomass per plant = C/B	Surivorship (%) per treatment = B/A (100)
2 seeds				
4 seeds				
8 seeds				
16 seeds				

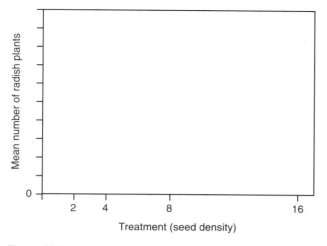

Figure 22-2

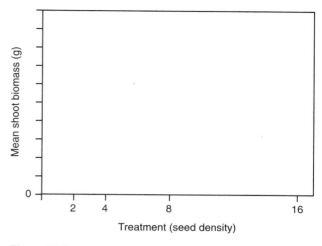

Figure 22-3

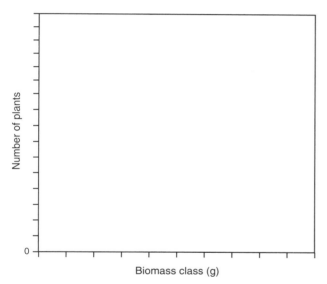

Figure 22-4

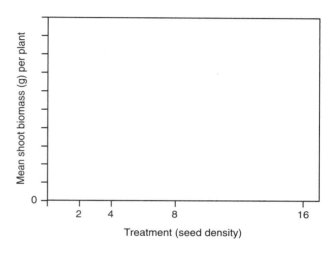

Figure 22-5

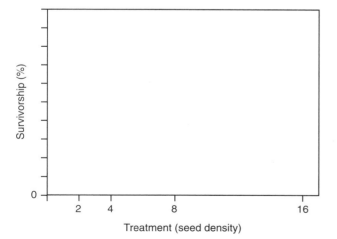

Figure 22-6

QUESTIONS FOR THOUGHT AND REVIEW

1. What is a "unitary" organism? _____

_____

_____

2. Give an example of a unitary organism.

_____

3. What is a "modular" organism? _____

_____

_____

4. Give an example of a modular organism.

_____

5. What is a population? _____

_____

_____

6. What is a rosette plant? _____

_____

_____

7. Define ecological competition._____

_____

_____

_____

8. What is meant by intraspecific competition?

_____

_____

_____

9. Why does the competition experiment begin with different numbers of starter seeds?

_____

_____

_____

10. What was measured at the end of the competition experiment? Why?_____

_____

_____

_____

EXERCISE 22
LABORATORY QUIZ

Name: _____

Section Number: _____

COMPETITION EXPERIMENT

1. What is meant by describing an organism as unitary? _____
_____

2. What is meant by describing an organism as modular? _____
_____

3. How would you determine population density for a population of unitary organisms? _____
_____
_____

For a population of modular organisms? _____
_____

4. What information was obtained by weighing the radish plants in the competition experiment?
_____
_____

5. In populations of unitary organisms, describe the type of size distribution you would expect to find.
_____
_____

6. Intraspecific competition means competition between _____
_____

7. Why did the size distribution of individual radish plants produce an L-shaped curve? _____
_____
_____

8. In scientific articles and reports, what is the difference between a results section and a discussion section?
_____
_____
_____